CLASSIC CARS AND
AUTOMOBILE
ENGINEERING

Volume 2

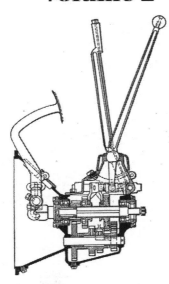

TRANSMISSIONS, AXLES, BRAKES,
WHEELS, TIRES, FORD CAR

the Kinneys

1

Classic Cars and Automobile Engineering: Volume 2

Transmissions, Axles, Brakes, Wheels, Tires, Ford Car

Restored by Mark Bussler

More books at
CGRpublishing.com

The Complete Ford Model T
Guide: Enlarged Illustrated
Special Edition

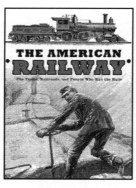

The American Railway:
The Trains, Railroads, and People
Who Ran the Rails

Antique Cars and Motor Vehicles:
Illustrated Guide to Operation,
Maintenance, and Repair

Automobile Engineering

A General Reference Work

FOR REPAIR MEN, CHAUFFEURS, AND OWNERS; COVERING THE CONSTRUCTION, CARE, AND REPAIR OF PLEASURE CARS, COMMERCIAL CARS, AND MOTORCYCLES, WITH SPECIAL ATTENTION TO IGNITION, START- ING, AND LIGHTING SYSTEMS, GARAGE EQUIPMENT, WELDING, FORD CONSTRUCTION AND REPAIR, AND OTHER REPAIR METHODS

Prepared by a Staff of

AUTOMOBILE EXPERTS, CONSULTING ENGINEERS, AND DESIGNERS OF THE HIGHEST PROFESSIONAL STANDING

Illustrated with over Fifteen Hundred Engravings

FIVE VOLUMES

LINCOLN SPORT PHAETON

4

Table of Contents

VOLUME II

Gasoline Automobiles

By Morris A. Hall   Revised by Tom C. Plumridge

PACKARD TWIN-SIX MOTOR SHOWING FUELIZER

TRANSMISSIONS

PART I

TRANSMISSION GROUP

Primarily, the clutch is used to allow the use of change-speed gearing; or, stated in the reverse way, the form of the transmission determines whether a clutch must be used or not, there being cases in which it is not used. Thus, where the frictional form of transmission is used, no clutch is necessary; the frictional discs act as a clutch and render another one superfluous. With the form of transmission known as the *planetary gear*, used in the Ford car exclusively, a clutch is used for high speed only.

On the other hand, the reverse of this does not always hold. Any form of clutch may be used with the various other forms of transmission, as the sliding gear; in fact, in actual practice every known kind of a clutch will be found coupled with the sliding-gear transmission.

Classification. Broadly considered, there are three classes of transmissions used. In cases where the use of any one of these forms eliminates the final drive, this from its very nature does not alter the facts but simply calls for a different and more detailed treatment. The three classes are: sliding gear operated in various ways, but usually selective; planetary, or epicyclic; and electric.

The features of the up-to-date transmission which stand out from previous years are: reduced sizes; simpler, lighter construction; greater compactness and greater accessibility. Perhaps the most noticeable trend has been toward the unit power plant which has helped materially to make transmissions smaller, lighter in weight, and more simple, with unusual compactness. This very compactness has brought with it a stiffness which has rendered less repairs and adjustments necessary, despite lighter weight. The smaller sizes have brought about the simplification and lighter weight, and in turn have been produced in answer to the popular demand for lighter weight cars. In part, simplification has been produced by unit power plants, now so popular.

Sliding Gears. *General Method of Operation.* Of the different types of sliding gears, the first two subdivisions are not very closely marked but blend somewhat into one another. The only real difference between them is the method of operation, the names serving to indicate the distinctive characteristics. Thus, in a selective gearset it is possible to "select" any one speed and change directly into it without going through any other, while, in the progressive form of transmission, the act of changing gears is a "progressive" one, from the lowest up to the highest, and *vice versa.*

Selective Type. With the selective method of changing gears, it is possible to make the change at once from any particular gear to the desired gear without passing through any other. Of course, the car will not start on the high gear any more than in the other case, but shifting into low for starting purposes is but a single action. So, too, when the car has been started, it can be allowed to attain quite a fair speed and the change to high made at once without going through the intermediate gears.

Progressive Type. Progressive gears, which are now little used, operate progressively: from first, or low, to second and from second to third, or high; in slowing down, from third to second to first and in this way only. This leads to a number of troublesome occurrences. In stopping it is necessary to gear down through all the higher speeds into low. If this is not done, when it is next desired to start the car, it will be necessary to start the engine, throw in the clutch, drop from the gear in mesh to the next lower, from that to the next, and so on down to low, throwing the clutch out and in for each change of speed. When first is reached, the car may be started. After starting, it is then necessary, in order to obtain any measurable speed with the car, to change back up the list, from low to second, from second to third, and so forth. In this way the progressive gear is disadvantageous, since its use means much gear shifting. However, the shifting is very easy for the novice to learn, as it is a continuous process, all in one direction.

Electrically Operated Transmission. In substance, the electrically operated transmission has all the hand levers, rods, and other levers replaced by a series of push buttons. When it is desired to change speeds, even before the actual change is necessary, the driver presses the button marked for the speed he thinks he will require. Then, when the actual need becomes apparent, he throws out the clutch and immediately drops it back again, all this forming but a single forward and back movement of the foot. During the slight interval while the clutch is out, the electrical connections shift

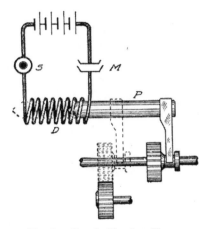

Fig. 1. Sketch Showing How a Solenoid Moves a Gear When Current Flows

the gears automatically, so that when the clutch is let back, the gears are meshed ready to drive.

To explain this action briefly, the gears are moved by means of solenoid magnets, which are nothing more than coils of wire through which an electric current from a convenient battery is allowed to pass. Through the center of each one of these coils passes an iron bar. When a current passes through the coil, it is converted into an electromagnet and draws the iron bar inward. As the other end of the bar is connected to the gear to be shifted, this movement of the bar shifts the gear. Consequently, when the button is pressed so that current flows through one of the coils, that action shifts the gear for which the button is marked.

The diagram, Fig. 1, shows but one pair of gears to be meshed, and the battery, push button S, coil D, iron bar P, and clutch connection M are all shown as simply as possible. When button S is pressed, current through the coil D will draw the bar P and mesh the gears as soon as the clutch has been thrown out, thereby closing the circuit at M. Fig. 2 shows the clutch pedal and its con-

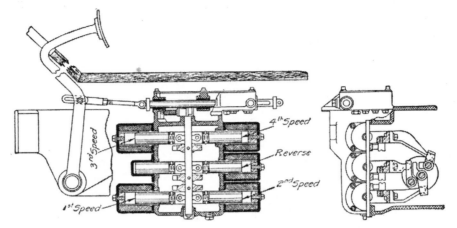

Fig. 2.   Arrangement of the Solenoids and Pedal in the C-H Electric Gear Shift

nection to the six solenoids necessary to produce four forward speeds, one reverse speed, and a neutral point.

On the steering wheel, Fig. 3, the control group of six buttons will be noted on the small round plate at the center, with the addi-

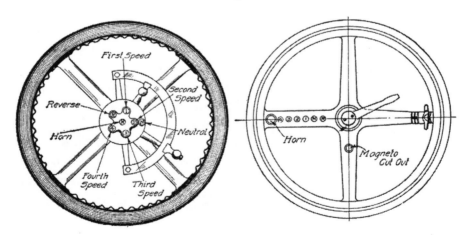

Fig. 3.   Arrangement of Buttons
for Gear Shifting

Fig. 4.   Another Arrangement of Buttons
for Gear Shifting

tion of the horn button in the center.   In Fig. 4 is another arrangement.

In the forms of electro-control systems, the buttons are grouped in one instance on the top of a small box four or five inches square, which is placed on the steering post below the wheel; in another, on the dash; and in a third, on a rod connecting post and dash.

Planetary or Epicyclic Gears. This type of transmission was used to a large extent in former years in such cars as the Cadillac and Winton, but it has lost in favor and is now used only in the cheaper cars. It is cheaper to manufacture than the sliding-gear type and easier to operate. The usual number of speeds is two forward and one reverse. The two types in general use are the internal gear and the spur gear, the latter being spoken of as the "all-spur-gear-type." The Ford transmission is of the spur-gear type and is by far the most successfully developed. There is a distinct advantage in the planetary one, because there is no power lost in friction when in high gear and the revolving parts tend to increase the flywheel effect and also makes the engine a great deal smoother in operation.

Method of Action. The principle upon which all planetary gears work is as follows: The first gear of the train is connected to the engine. The second is one of a series of several gears; these are pivoted in a drum, which may be held stationary by a brake band. The middle, or third, gear in the train, as well as the last, or fourth, is connected to another gear, *a driven gear, not a driver.* Considering but a single rotating train—there usually are three or more—the last-named gears form the fifth and sixth in the whole train. Gears two, three, and four have different numbers of teeth, as well as gears one, five, and six. Holding the band which holds the drum to which the gears are pivoted allows each of them to rotate around its own axis but not around the main shaft. This form of rotation gives one gear reduction.

Another band holds another gear stationary and allows the three-gear unit to rotate around the main shaft as an axis; at the same time it leaves them free to also rotate around their own axis. This produces another gear reduction. Another form which is popular in so far as planetary gears are popular is that in which internal gears are substituted for one set of the planets, from which the device obtained its name. This does not complicate the device any; in fact, the only way in which it makes any change is in the manufacturing cost of the gear, internals costing more than spur gears.

Ford Planetary Type. Ford has been a consistent user of the planetary gear; in fact, the simplicity and ease of operation of his well-known and widely used car is largely due to this use. The Ford transmission, which is of the all-spur-gear type, is shown in Fig. 5. This is operated by means of two pedals and a lever, one pedal

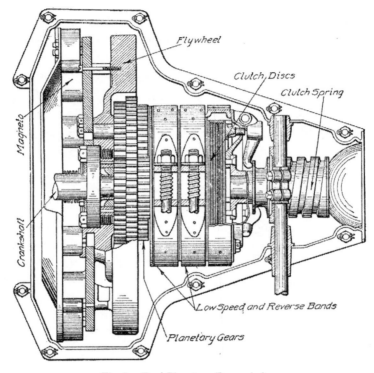

Fig. 5. Ford Planetary Transmission

working high and low speeds, while the other pedal controls the reverse. The first-named pedal, however, must be used in conjunction with the forward movement of the hand lever which locks the high-speed clutch, seen in this figure at the right.

MISCELLANEOUS TYPES

Freak Drives. What are termed the freak drives attract much attention from inventors and but little from hard-headed constructors. Thus, the belt drive was once advanced as the simple drive, yet it made no progress. Today there are few belt drives used in final driving in America, although a few are still made in Europe.

There is a low-price French car, Fouillaron, with this drive; and a single-cylinder Italian car, the Otav, selling for the equivalent of $150, which is also equipped with a belt drive.

Cable and Rope Drives. When cycle cars were first brought out and by many considered as destined to replace both the low-priced cars, on account of their still lower price and simplicity, and motorcycles, because of their greater comfort, superior appearance, and greater carrying capacity, many of the simple drives were revived and applied to the cycle cars. The types used include the cable drive, which attracted much attention at one time in the motor-buggy field, the rope drive, the flat belt, the V-belt, the cloth-covered chain, and many others. With the collapse of the cycle-car boom, these went out of use.

Hydraulic Gear. The hydraulic transmission has been advanced as a cure for all automobile troubles, representing as it does the elimination of clutch, differential, and the driving mechanism. It consists of a pump to circulate the fluid, and one or two motors usually attached to the rear wheels and propelled by the fluid.

Transmission Location. There are but four recognized positions for the transmission in the modern car—(1) unit with the engine (unit power plant), (2) amidships in unit with clutch or alone in a forward position, (3) amidships in unit with forward end driving shaft or in a rear position, and (4) at the rear in unit with rear axle. The use of the rear axle and the transmission combined into one unit is rapidly dying out. There is only one make of car and only one model now using this type of installation.

Unit with Engine. The single unit, which is the combining of the transmission and engine, is the most popular of installations at the present time and gives greater accessibility for repairs as well as stronger and more rigid construction. It also gives perfect alignment of all shafts and other operating parts as well as compactness and lightness. A notable example of this type of installation is shown in Fig. 6, which is the Packard "Straight-Eight" engine, in which the cylinders are all in line. The transmission is shown at the rear end of the engine and attached to a bell housing which is a part of the engine crankcase.

Another example of the unit with engine type is seen in the Grant-Lee three-speed gear box, Fig. 7, as utilized in the Hackett

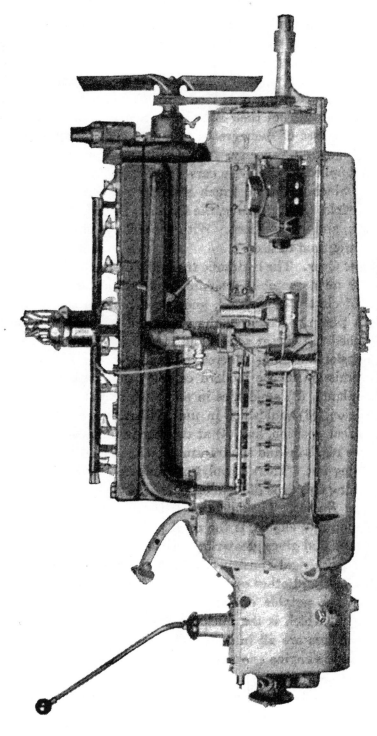

Fig. 6. Packard Transmission in Unit with Engine
Courtesy of Packard Motor Company

car. This is unusually small and compact, as will be noted by comparing the size of the unit with the operating levers and pedal. While the clutch is not shown, its housing is, also the flange which attaches it to the flywheel housing to complete the power unit. A third example of the engine-unit power group is shown in Fig. 8, which shows the flywheel, clutch, and transmission of the Peerless eight.

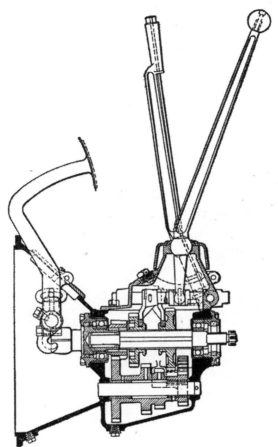

Fig. 7. Gear Box Used in Hackett Cars Is Very Small and Compact
Courtesy of Hackett Motor Car Company, Jackson, Michigan

This unit transmits many times the power of the Hackett unit and is therefore much larger. In this unit the bearing arrangement is rather unusual, as roller bearings of the taper form are used on the main shaft, a straight roller for the spigot bearing, and plain ball bearings for the countershaft. The shortness and large diameter of the shafts should be noted.

GASOLINE AUTOMOBILES

Amidships Alone or with Clutch. This type of installation was used when the transmission was heavy. It had wider use in Europe than in this country. The Winton used this type of installation, but in general it is not a common construction on pleasure cars, although it is used on quite a number of trucks. On the amidships-

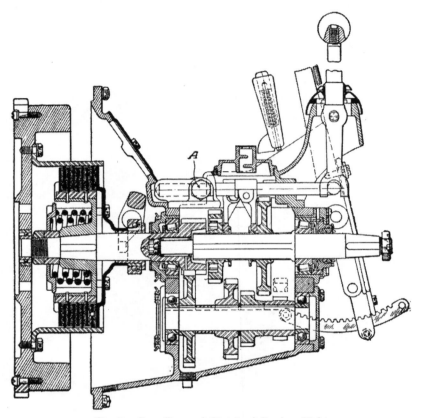

Fig. 8. Gear Box and Clutch of Peerless Eight
Courtesy of Peerless Motor Car Company, Cleveland

clutch unit type, however, the combination is not quite so intimate as the one in which the two units are enclosed in a common case.

Amidships Joined with Driving Shaft. The amidships unit joined with the forward end of the driving shaft is shown in Figs. 9 and 10. The universal joint with the driving-shaft pivots is seen at the left side of both these views. In this construction there is usually a frame cross-member at the point on which the rear end of the transmission is supported. This same arrangement is used on

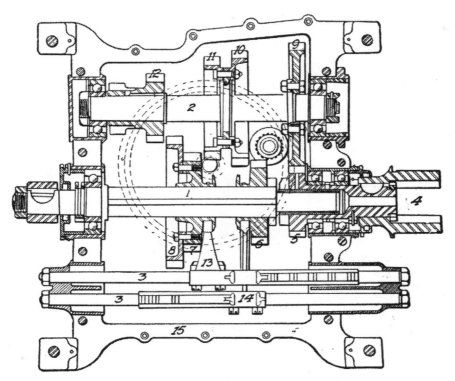

Fig. 9. Sectional Plan Drawing of the Locomobile Four-Speed Transmission
Courtesy of Locomobile Company of America, Bridgeport, Connecticut

Fig. 10. Photographic Reproduction of Locomobile Gear Box Shown
in Section in Fig. 9

the Stearns-Knight four-cylinder chassis, the transmission of which is shown in Fig. 11. In this transmission, the stiffness of the cross-member at the rear end of the transmission is also utilized to support the brake drum of the foot-brake system. The short stiff shafts on the transmission will be noted, also the many splines on the main shaft and the use of double row ball bearings on the main shaft, with

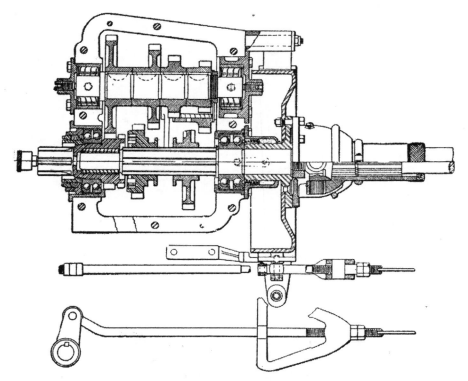

Fig. 11. Three-Speed Transmission and Brake of Stearns Four-Cylinder Car
Courtesy of F. B. Stearns Company, Cleveland, Ohio

a flexible roller on the spigot, and the same type of bearings on either end of the layshaft, which is alongside of the main shaft. Note also the means provided for adjusting the countershaft longitudinally by the two steel screws projecting through the bearing caps so that this adjustment can be made from the outside.

Unit with Rear Axle. The position of the unit at the rear axle is not so widely used as a few years ago. The Stutz car has a typical rear-unit construction and is about the only car which uses this equipment. The Studebaker makers used this construction for a number of years, but they are now mounting the transmission amidships. The Stutz rear axle and transmission are shown in Fig. 12. There are two brake shoes: one for the service brake and the other for the emergency brake. The position of the transmission calls for two operating rods, each the full length from the operating levers to the rear axle.

The connections of the levers and the shifting of the gears are plainly shown in the Studebaker transmission, Fig. 13. The gear-

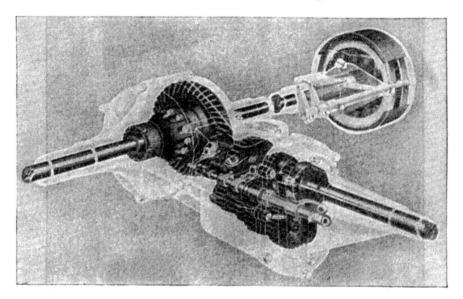

Fig. 12. Stutz Transmission Mounted in a Unit with the Rear Axle

shifting lever is placed in the center and is shown solid in the neutral position and lighter in the other four positions. Just below these levers the transmission is shown with the position of the gears for neutral. The high and intermediate sliding gear is between the countershaft gear and the intermediate gear, and the low and reverse sliding gear is between the low-speed countershaft gear and the reverse countershaft gear. The four corners show the positions of the gears when the lever is in each one of the positions.

These positions plainly indicate that there is a driving gear and two slide members on the main shaft and four gears on the countershaft. At the left in the picture, the gear toward the rear is for reverse. An idling gear (not shown) is needed to complete

the reversal of the motion. When the lever is swung to the left
and forward, this group is completed and the speed is reversed.

At the time when the shifting lever is swung to the left and
backward, the rear sliding member is moved forward to mesh with
the second gear on the countershaft as is shown in the diagram,

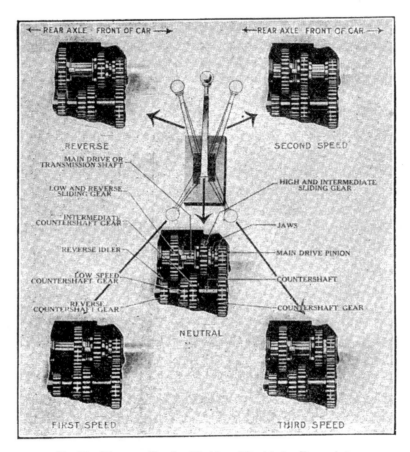

Fig. 13.   Diagrams Showing Working of Studebaker Transmission

Fig. 13, at the lower left-hand corner, and first, or low, speed results.
With this gear in neutral position—as it is left when the shifting
lever is swung through the neutral position and over to the right—
a further movement to the front picks up the forward sliding gear
and moves it back into mesh with the third gear on the layshaft.
This combination, as shown in the upper right-hand corner, gives
second speed. When the lever is moved back, it moves the gear

forward, giving high speed and direct drive, as shown at the lower right-hand corner.

Modern Transmission Construction. The transmission and clutch unit of the LaFayette car is shown in Fig. 14. The clutch

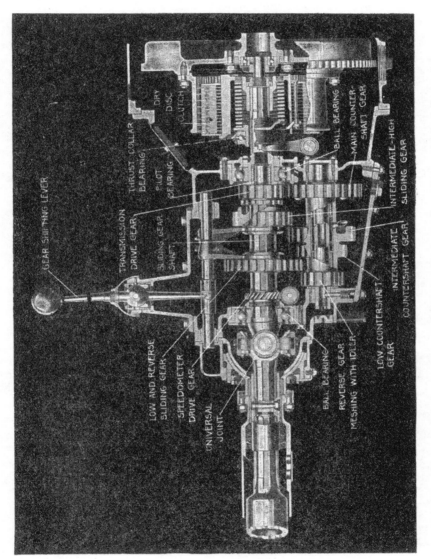

Fig. 14. Transmission and Clutch Unit of the 1920 LaFayette Car. Universal Joint is Enclosed in the Transmission Case

is a multiple-disc dry-plate type, having a fabric lining riveted on the driving or larger discs. The multiple-disc clutch is the most popular form now in use and the majority of the present-day cars are being equipped with it. The clutch spring is made of coiled flat wire and the thrust of this spring is taken up by a ball-thrust bearing. Pressure on this thrust bearing is had only when the clutch is thrown out or disengaged. This bearing should be lubricated occasionally so that it will not become noisy. If the clutch pedal is adjusted so that there is continual pressure on this bearing, it will wear out very rapidly. The dry disc plates will last for an indefinite time; sometimes, however, they become glazed and it is then necessary to soften up the fabric surface either with emery cloth or a little oil.

The transmission furnishes three speeds forward and one reverse. The majority of cars now have this speed arrangement. There are two gears on the sliding gear shaft—the low and reverse sliding gear and the intermediate and high sliding gear—that are placed in mesh with the proper countershaft gears when a speed change is desired. The sliding gear shaft is mounted at the front end in a pilot bearing of the roller type. Bronze bushings were generally used at this place, but they were unsatisfactory as they wore out in a short time. This pilot bearing is placed within the clutch shaft and this shaft is supported in the transmission case by a ball bearing. The countershaft is stationary in the case and is hollow to permit oil being conducted through a port at the front end of this shaft, through the shaft and to the universal joint compartment. This keeps the oil level in the universal joint housing a little above that in the transmission case.

A great number of transmissions have bearings on each end of the countershaft and the shaft revolves. There is a disadvantage, however, in this form of construction as there are likely to be oil leaks at the ends of the shaft. The transmission drive gear is constantly in mesh with the main countershaft gear, causing the countershaft gear assembly to revolve whenever the motor is running. This continuous operation wears the roller bearings but very little as there is no power being transmitted through these gears unless one of the sliding gears is in mesh with one of the countershaft gears. The inside of the countershaft does not communicate with the roller bearings. These bearings, however, are lubricated by oil picked up by the scoop shown near the intermediate countershaft gear in Fig. 14.

The speedometer is driven by the gear that is mounted on the rear end of the sliding gear shaft, which is of the spiral-cut type.

It drives a worm gear that is attached to the speedometer drive cable. These two gears are often placed inside a housing just outside the main transmission case and ahead of the propeller shaft.

Modern Selective Types. The various differences, advantages, and disadvantages of the modern selective types of gear boxes may be seen in Fig. 15. This type shows the three-speed selective gear used on the Cadillac cars, which is but slightly modified from the type in use for several years. This change should be noted, however;

Fig. 15. Cadillac Transmission and Housing

the layshaft, which formerly was on the same horizontal level as the main shaft, is now placed directly below it. This makes a higher but narrower gear box, that is, instead of being wide and fairly flat, it is now high and narrow. The placing of the shifting levers on the cover, directly over the center, has aided in making the gear set more compact than formerly. There is a tendency toward the elimination of the change-speed lever and emergency brake lever from the driver's compartment. In the Apperson car the shifting of the gear is done mechanically and the selector lever is placed on the steering column. There are two shifting gears in the column, one gear carrying a set of dogs cut into its face, which mesh with a similar set on the main driving gear to give the direct drive. The gear portion of this member meshes with another gear for second. The second shifting member meshes with one gear on the layshaft for low speed and with another on the third shaft for reverse. The reverse gear is at all times in mesh with the fourth layshaft gear, so that on reverse the drive is through five gears instead of four. On high gear the drive is through the dogs, the layshaft being driven, of course, but silently, as it transmits no power.

Four-Speed Type with Direct Drive on High. One of the tendencies of recent years has been the gradual change toward more speeds, as shown by the increasing use of four-speed gear boxes. Other indications of this change have been the two-speed axle, which gave double the number of gear-box speeds, with the ordinary three-speed and reverse transmission; and the electric transmission, which affords seven forward and two reverse speeds.

Following this increase of speeds, the multi-cylinder motors and downward price revisions of the early part of 1916 brought about a combination which almost eliminated the four-speed gear box or at least removed it from all but the most expensive of cars and from many of those. It is claimed that the eight- and twelve-cylinder motors have so much power and flexibility that a fourth speed is rendered unnecessary. The four-speed gear box is more expensive than the three-speed box, and the lowered prices of cars have been instrumental in preventing its continued use. At the same time, there was considerable lightening of weight all over the chassis, and the four-speed gear box had to go out of all but the biggest cars on account of its greater weight.

One of the few four-speed gear boxes left is shown in Fig. 9. The two-gear shafts, as well as the operating shafts, lie in the same horizontal plant. Fig. 10 shows the location of the shafts more plainly. Both forms show the arms which project up to attach this unit to the frame. The cover, which is a light easily removed aluminum member, is taken off from above after the floor boards are lifted out. This arrangement makes for accessibility and eliminates any need for lying on the ground while working on the transmission gears or shafts.

GASOLINE AUTOMOBILES

The form of final drive alters the construction of the transmission very materially. Formerly, when all final drives were of the double-chain form, it was customary to include the differential, bevel gears, and driving shafts in the gear box. Now that the chain has gone out, this construction is found only when the gear box is a unit with the rear axle.

Four-Speed Type with Direct Drive on Third. In all the transmissions described thus far, the direct drive has been the highest speed and is used because it gives quieter operation in traffic. There are many makes that use the direct drive on third speed, for example, the Winton; the fourth being a geared-up speed for use only in emergencies, when the very highest rate of travel is required, and when a little noise more or less would make no difference. This arrangement of the direct drive and silent speed has long been a debated point, some designers favoring the direct-drive type with an over-geared speed for occasional use, while the opponents of this method say that this construction practically reduces the transmission to a three-speed basis, the fourth being so seldom used that it is practically negligible. They say, also, that the modern motor can attain a high enough speed, on the one hand, and is flexible enough, on the other, to permit its being used with the high-gear direct drive upon almost all occasions.

Constant Mesh Gear Transmission. An excellent example of this type of transmission is the Campbell unit installed on the Chandler car. It is a distinct departure from the usual method of gear change and gear arrangement. The gears are continually in mesh and there is no movement of the gears when a speed or gear change is made. In most all transmissions, the gears are moved along the shaft when a change of gear is required, but in the Campbell transmission, a gear change is made by moving keys which slide in slots in the main drive shaft of the transmission. In the sliding gear transmission, the gear and shaft revolve together, while in the Campbell, the gears are on collars which revolve around the shaft until the key for that gear is moved into position at which time gear and shaft revolve as one unit. These keys are of peculiar shape and are shown in Fig. 16, and the knife-like edge takes the thrust and drive. Before the key is moved, it is in the position shown in Fig. 17 and as the key is moved into position, a spring plunger forces the

Fig. 16.   Rocking Keys
Courtesy of Campbell Transmission Company and Chandler Motor Car Company

CHANDLER TRAFFIC TRANSMISSION

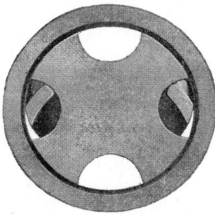

DIAGRAMMATIC VIEW WITH ROCKING KEYS IN NEUTRAL POSITION UNDER BEARING RING.

Fig. 17.   Keys in Neutral Position
Courtesy of Campbell Transmission Company and Chandler Motor Car Company

CHANDLER TRAFFIC TRANSMISSION

DIAGRAMMATIC VIEW WITH ROCKING KEYS ENGAGED IN INTERNAL NOTCHES OF GEAR

Fig. 18.   Keys in Driving Position
Courtesy of Campbell Transmission Company and Chandler Motor Car Company

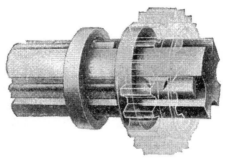

Fig. 19.   Key in Position Under Gear
Courtesy of Campbell Transmission Company and Chandler Motor Car Company

key to roll into the driving position as shown in Fig. 18. When the key is in position, one side of it takes the drive while the other side takes up the back-lash and keeps the gear tight in position. Fig. 19 shows the key in position under a phantom gear, while Fig. 20 shows all parts of this transmission. The operating plungers are shown towards the rear of the transmission. At one end of the sliding gears

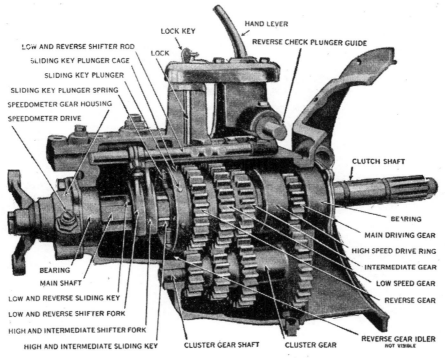

Fig. 20. Parts of Campbell Transmission
Courtesy of Campbell Transmission Company and Chandler Motor Car Company

is a groove in which a collar revolves. This collar moves the key and enables it to be shifted when the shafts are revolving. The advantage of this constant mesh-gear transmission is that the gear change can be made to a higher or a lower gear without any clashing of gears or noise of any kind. It is a great factor for safety in hilly country because the gear can be changed to a lower gear at any time and in this way the engine can be used as a brake.

There are other types of constant mesh gears in which dog clutches are moved along the shaft to mesh with other dogs attached to the gears, but the Campbell is the only one that uses the sliding key. The principle of the sliding key is not a new one, since it was used in punch presses for many years. It was found to stand up under this heavy work and can be fully relied upon in the automobile transmission. A key has never been known to break or shear under ordinary operating conditions.

TRANSMISSIONS

PART II

Interlocking Devices. Nearly all transmissions have a form of stop lock on the shifting rods in the transmission, which holds the gears in mesh until they are moved again. In reality this arrangement simply prevents the gears from jumping out of mesh and

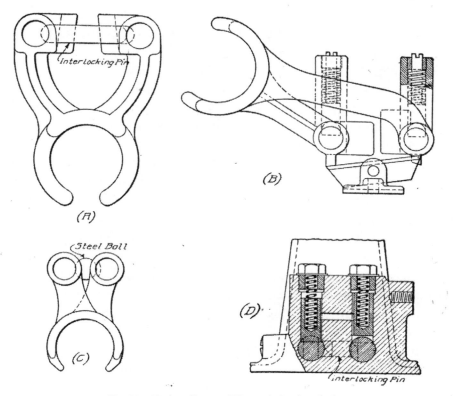

Fig. 21. Various Forms of Transmission Interlocks

generally, the most simple arrangement which will hold the gears is used. In the ordinary form, this arrangement consists of hardened steel wedges with light springs back of them and deep grooves in the shifting rods into which these wedges fit.

Not all transmissions have the wedge and notch. In Fig. 21 a method of interlocking by means of a pin at the shifting forks (not rods) which project into shallow holes in the two shifters is shown at *A*; a rocking, or tilting, bar beneath the shifting forks, which is pressed into a notch in either fork when moved from neutral, is shown at *B*; and at *C*, the use of a steel ball—all three arrangements being used by the American Die and Tool Company. The form at *D* shows the pin used by Grant-Lee Gear Company, the grooves in the rods being deep enough to accommodate this form in a neutral position so that the rod can be started. But the guide hole in the central housing in which the pin is moved across by the motion of one rod, owing to the shape of the bottom of the groove, prevents the other rod from moving.

Pneumatic Shifting System. The pneumatic system of gear shifting is along lines somewhat similar to the electric system; air under pressure being used to move the gears instead of a hand lever and rod combination. For this purpose it is necessary to add an air compressor, a tank to carry the compressed air, and what is called the "shift"—really a complicated valve and a series of plungers —to the car. The valve and plungers respond to a finger lever on the steering wheel, the same as the electric system responds to the buttons. Air is admitted behind the plungers, which moves the gears as soon as the clutch is depressed. This system, like the electric shifter, permits the anticipation of the needs of the car.

Railway Car Needs. All transmissions previously presented have had but one reverse. For gasoline railway cars, the inability to turn the car requires as many reverse speeds as forward which means special gearing. Usually, this gearing is accomplished by means of a pair of bevels, each with a clutch, meshing with a single driving wheel. Obviously the two driven bevels will turn in different directions, and each will drive when its clutch is engaged.

Transmission Operation. Practically all transmissions operate the gears by means of a long hand lever, placed either at the side of the car or in the center, according to the location of the control. Even on planetary forms at least one of the various speeds is controlled by a hand lever. The electric- and air-shifting methods have made a good start, but until their number increases materially, these types can be considered as only having started their development.

Transmission Lubrication. A fairly heavy lubricant is generally recommended for gear-box use—either a special form of about the right consistency, or else a home-made mixture of about half-and-half of light oil and hard grease. Some firms recommend a graphite grease. The lower part of the case should be filled to a point, or level, where the largest gears dip continuously. This will insure a constant agitation of the lubricant, which will thus get to all moving parts and surfaces. If the lubricant is too stiff, the gears simply cut a path through it. This results in all other parts running practically dry. Too thick a lubricant or too much of it will make a fairly heavy drag on the motor, which loss of power should be avoided. Gear-box lubricant generally is introduced in bulk by the removal of the cover. The outside parts carry their own grease in cups.

Transmission Bearings. By looking back at the various transmissions shown, it will be noted that ball bearings have the widest use. Roller bearings in various forms are coming into use, as the shorter series produced in the last couple of years has shown designers that this type would produce a compact gear-box, their size having previously limited their use. Plain bearings are not used on good cars.

Transmission Adjustments. Few adjustments are needed in the modern gear box. However, provision for wear is made in the operating rods and levers, both within the case and without. In some cases, the shafts may be slightly shifted endwise to secure better meshing of the gears after wear. Bearings, too, are arranged to shift slightly in an endwise direction to take care of wear in other parts.

TROUBLES AND REPAIRS

Noise in Gear Operation. One of the most common of transmission troubles is a grinding noise in the operation of the gears. This is heard more in bevels than in spurs, but in old transmissions and on the lower speeds it is heard frequently. A good way to quiet old gears, after making sure that they are adjusted and meshing correctly, is to use a thicker lubricant. If thick oil is being used, change to half-oil and half-grease or preferably all grease.

Another source of gear-set noise is a shaft out of alignment, caused either by faulty setting, by worn or loose bearings, or by yielding or cracking of the case. If it is properly set at one end and is out at the other, the trouble will be more difficult to find and remedy.

Worn Bearings. There are two types of bearings used in transmissions—the ball bearing and the roller type. If there is a great deal of play in either type, the transmission will be noisy in operation. It is therefore essential that the bearings be kept in proper adjustment at all times. If the bearings are operated when loose or out of adjustment, they will be damaged to such an extent that they will have to be renewed; whereas, if adjusted as occasion demands, they will last indefinitely. In the annular or ball bearing there is usually no adjustment, while there is one in the roller type. After a period of service, end play in the transmission shafts will develop and this should be taken up at once. End play is the lateral movement of the shaft and can be found by pushing or pulling on the shaft to be adjusted. The adjustment is correct when there is no noticeable end play and the shaft can be easily rotated by hand. The adjustment is made by moving either the inner or outer ring of the bearing. Some of the adjustments as used on the Brown-Lipe transmissions are shown in Fig. 22. The adjusting part of the bearing is moved by a threaded nut which is locked into place by a pawl or other device which fits into slots cut in the adjustment unit. An adjustable transmission can be easily recognized because the adjustment nut is on the outside of the transmission case. To see if there is any play on the countershaft, a pry or large screw driver should be used against a gear in one direction and then another. If the transmission is being overhauled, the bearings should be examined, whether they are roller or ball bearing. The races of the ball bearings are sometimes found to be pitted or rough and this condition causes a rumbling sound in the bearing. In the roller type, grooves will sometimes be worn, especially if the bearing adjustment has been neglected and this condition will also cause a rumbling. If too much end play is allowed, the gears will not mesh properly and will cause uneven wear on the gears as well as giving the gear a chance to jump out of mesh where it is doing the work of driving the car. Play in the transmission bearings can only be found and avoided by strict periodical inspection and attention given at once will give prolonged life to every part of the unit concerned.

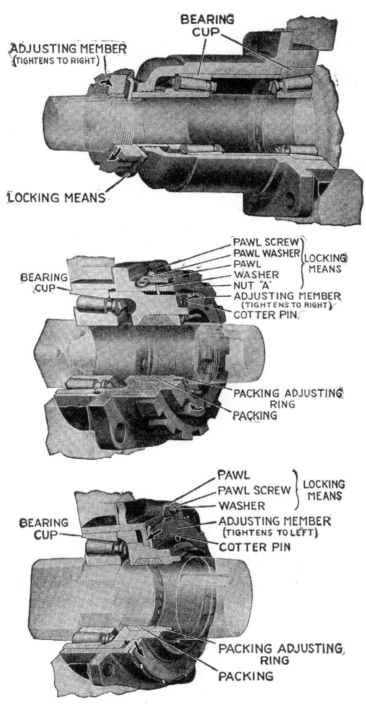

Fig. 22. Adjustments as Used on Brown-Lipe Transmissions
Courtesy of Brown-Lipe Transmission Company

Gears Jumping Out of Mesh. In the case of the high gear, this trouble can nearly always be laid to worn gears and is usually in evidence by the gear jumping out of mesh when the car is pulling hard. The tips of the high speed gear, that is, part of the clutch

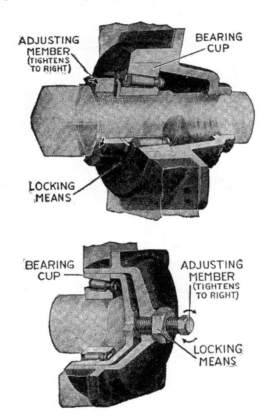

shaft, wear off and throw the sliding gear out of mesh. This trouble can also be caused by the gear-locking device springs becoming weak or the slot in which the plungers fit becoming worn and allowing the gears to slip out of mesh. The worn parts should be replaced.

Heating. Heating is a common trouble, and can usually be traced to lack of lubricant in an old car or to too large shafts or too small bearings in a new one. Sometimes the grease used will cause heating, particularly when long runs are made with the transmission working hard. This is most noticeable when the grease or lubricant is of such a consistency that the gears simply cut holes in it but do not carry any around with them or do not otherwise circulate the lubricant. This can be remedied by making it thicker so the gears will cut it better, by making it thinner so they will splash it more, or by changing the nature of it entirely.

Gear Pullers. One of the principal necessities for transmission work is a form of gear puller. These are like wheel pullers, except

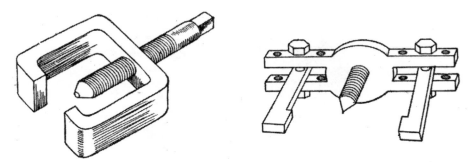

Fig. 23. Types of Gear Pullers

that they are smaller and more compact. In Fig. 23, a pair of gear pullers are shown. The one at the left is very simple, consisting of a heavy square bar of iron which has been bent to form a modified U. Then a heavy bolt is threaded into the back of this, or bottom of the U. This will be useful only on gears which are small enough to go in between the two sides of the puller, that is, between the sides of the U, which when in use is slipped over the gear, the screw turned until it touches something solid, as the end of the gear shaft, and then the turning continued until the gear is forced off.

While not as simple as this, the form shown at the right has the advantages of handling much larger gears and also of being adjustable. It consists of a central member having slotted ends in which a pair of L-shaped ends, or hooks, are held by a pair of through bolts. Then there is a central working screw. To use, the hooks are set far enough apart to go over the gear, then slipped around it and hooked on the back. The central screw is turned up to the end of the shaft, and then the turning continued until the gear comes off. There are many modifications of these two; in fact, practically every repair shop in the land has its own way of making gear or wheel pullers.

Pressing Gears on Shafts. The opposite of pulling off gears is putting them on. Very often they are designed to be a press fit, which means exerting tremendous pressure. Every repair shop should have some form of press for this and similar work. In Fig. 24 the man is just beginning to apply pressure to the shaft to force it into the lower gear. The table must be arranged for work of this kind with a solid spot when the shaft does not come through, and

Fig. 24. Method of Pressing Transmission Gear onto Its Shaft

with a hole when it does. The work of pressing is usually done in a few seconds, while the preparation, alignment, and starting of the work takes perhaps half an hour or more. It is work which should be done very carefully.

One way in which arrangement can be made for pressing a shaft a considerable distance into a gear and, conversely, for pressing the shaft out of the gear, is that shown in Fig. 25. This press has the additional advantages of being simple, easily constructed, and cheap. A solid base is constructed with a pair of hinged uprights. These can be dropped together with the work between them, forming a modified triangle, the strongest known shape, resting upon its broadest side and thus having the greatest stability. With this arrangement the press can readily be used for pressing off parts.

Care in Diagnosis. The repair man should use a great deal of care in diagnosing the trouble in a transmission. Frequently, what appears at first to be at fault turns out to be all right, and something else is back of the first trouble, which must be corrected before a remedy can be applied. For example, a repair man may figure that a new gear is needed to repair a transmission. The gear is received from the factory three days later, but when he starts to put it in, he finds that a bearing is defective; in fact, the defective bearing caused the wear in the gear. This necessitates a further delay of three days in order to get a new bearing.

Poor Gear Shifting. A common transmission trouble is poor gear shifting. This may be due to a number of different things.

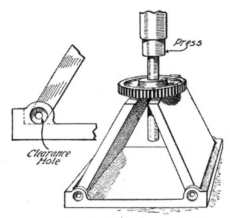

Fig. 25. Home-Made Table for Use in Gear Pressing or Pulling
Courtesy of "Motor World"

The edges of the gears may be burred so that the edges prevent easy meshing. Any attempt to force the gears into mesh only burrs up more metal and makes the situation worse. Whether this is the trouble or not can be determined very quickly and easily by removing the transmission cover and feeling of the gears with the bare hand; the burred edges can readily be distinguished. If this is the only fault, the transmission should be taken down, the gears taken out and placed in a vise, and the burrs removed with a cold chisel and file.

Poor or worn bearings or a bent shaft or one not accurately machined may cause difficult shifting. If the bearings are worn, the difficulty of shifting will be accompanied by much noise, both in shifting and after. The bent shaft is more difficult to find and equally difficult to fix. A new shaft is usually the quickest and easiest way to remedy the trouble.

Sometimes the control rods or levers bind or stick so that the shifting is very difficult. In case the gears are difficult to "find" or will not stay in mesh, the fault may be in the shifter rod in the transmission case. This usually has notches to correspond to the various gear positions, with a steel wedge held down into these notches by means of a spring. The spring may have weakened,

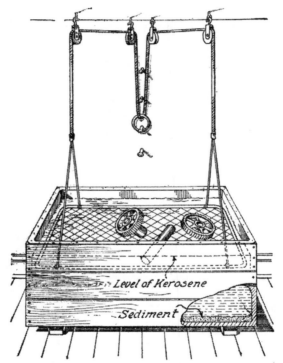

Fig. 26. Tank and Basket for Cleaning Gears and Other Parts
Courtesy of "Motor World"

may have lost its temper, may have broken, or for some other reason failed to work. If the springs are in good working condition, the edges of the grooves or notches may have worn to such an extent as to let the wedge slip out of, or over, them readily.

Cleaning Transmission Gears. When the transmission is taken out of the case and has to be taken apart, and particularly if it has not been cleaned for a long time, it is advisable to clean all the parts thoroughly before attempting to work with them. The best way to clean the parts is to have a special cleaning tank, Fig. 26, which is not unlike the baskets used in some hardening processes. It consists of a deep metal or metal-lined tank and a basket or tray, which is an easy fit in it, suspended from above by wire cables. The cables are brought together on the wall, where a ring joining the ends and a series of nails or hooks make it easy to hold it at any desired eleva-

Fig. 27.   Handy Framework for Lifting Transmissions out of Chassis

tion, either in or out of the tank. The tank is filled preferably with kerosene. As soon as a part has been removed from the transmission, it is thrown into the basket, and when this is filled or all the parts are in it, it is lowered into the kerosene and allowed to stand, for a couple of hours if possible. When thoroughly soaked, the basket should be raised above the level of the liquid and allowed to drain thoroughly. If it can be left for an hour or so to drain, all trace of kerosene will disappear and the gears, shafts and other parts will be like new.

Lifting Out Transmissions. When the trouble has been found to be in the transmission case or in some part that necessitates complete removal, it is often a tremendous job to get the unit out. Units attached from below are not so difficult to detach. They are lowered by means of a platform of boards set on two or more jacks. But when a unit must be removed from above and no overhead beam is available, the hoist shown in Fig. 27 will be found very handy. This hoist is simply a triangular framework constructed from angle iron to have the minimum height which will allow removal of the unit. The chain fall is attached to a hook in the center, and the

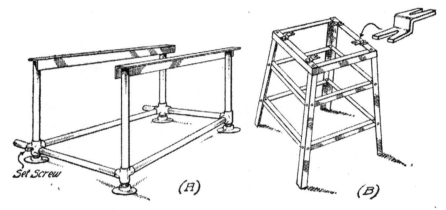

Fig. 28.   Two Forms of Useful Transmission Stands
Courtesy of "Motor World"

chains put around the case. When lifted up close into the V of the framework, the whole transmission can be put onto horses and moved along the chassis, or boards can be put under it and over the chassis frame to allow it to be worked there. If desired, it can be lowered onto a creeper or other low platform with wheels and moved out of the way. This rigging can be used for many other similar purposes, although it is not suitable for the removal of an engine, radiator, or other part or unit which extends far above the chassis frame.

Transmission Stands. When the transmission has been removed, if the work to be done upon it is all extended, a stand to support it is really a necessity if the work is to be done right. A pair of stands are shown in Fig. 28, the one at the left is made from pipe fittings and angle irons in such a way that the width between the rails can be varied to suit the transmission or engine. The stand at the right is more of a specialized type. It is constructed for a certain transmission and has clips to support it in the same way that it is held in the chassis. The latter frame may be smaller and more compact than the former, but the wide range of uses to which the former can be put make it more desirable in the average shop.

"Working in" Bearings. When a great many bearings of any one transmission are fitted, it is well to make a jig for "working in" the cases to an exact size for the bearings, whether these be over

Fig. 29. Method of Fitting Transmission Case Bearings with Dummy

sizes or not. Fig. 29 shows an aluminum transmission case with a pair of jigs for scraping its bearings into the case. These jigs are made of steel and are constructed to a very accurate size, the surfaces being hardened so they will show no wear. The jigs are painted with Prussian blue, put in place and turned, the markings scraped by hand, the jigs again put in place and turned, and this process repeated until a perfect bearing surface is obtained. Starting with an unknown size on the case and a known size of bearing which must go in it, a few of these jigs will soon save their cost in labor and time by quickly producing the necessary size of case to take the bearings.

Saving the Balls. If a great many ball bearings, particularly from transmissions, are used, and many bearings scrapped, it is advisable to save the balls. These balls will come in handy later for replacement or other uses. Moreover, balls are expensive, and good ones are hard to obtain. A handy way to take care of balls without much work beyond cleaning thoroughly in the kerosene tank is to construct a cabinet like the one shown in Fig. 30. There are four drawers—or more if desired. The bottom of each drawer is a steel plate drilled as full of holes as possible of the next smaller size, that is, a clearance size for the next round figure size. Then the

Fig. 30. Easy Way of Sorting and Keeping Old
Bearing Balls
Courtesy of "Motor World"

cabinet does the sorting, all balls being put into the top drawer. The next smaller size is retained in the second drawer, the third size in the next, and so on. When using balls out of these drawers, the micrometer should be used to determine their exact size.

Handy Spring Tool. In the Ford transmission-band assembly there are three springs which it is difficult to assemble because of the trouble in holding so many things at once. To eliminate this trouble, the tool, shown in Fig. 31, made from flat bar stock, can be constructed. The handles, if they could be called that, are pivoted together and carry a kind of flat jaw with three notches at one end. When the two of these are squeezed together by means of the screw and handle at the other end, the flat plates will hold the three springs tightly enough so that all can be inserted in their proper positions at once by using but one hand. Tools of this kind, which save a great deal of the workman's time and thus save both time and money for the owner of the car, should, and in fact do, distinguish the well-equipped repair shop and garage.

In transmissions of the planetary type, there is little or no trouble except with the bands. If these are loose, the gears will not engage and the desired speed will not result. If they become soaked with grease, oil, or water, they will not work as well as if kept clean and, in the case of excessive grease, will slip continually. If the band lining becomes worn, it should be treated just as a brake lining is. When inspected for wear and found not badly worn but slippery, it may be cleaned in gasoline and then in kerosene, after

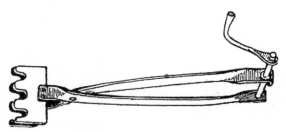

Fig. 31. Handy Spring Tool for Ford Assembly

which a saw, hacksaw, or coarse file may be used to roughen it. Sometimes greasy bands can be fixed temporarily—say, enough to get the car to a place where tools, materials, and facilities for doing the work are available—by sprinkling them with powdered rosin or fuller's earth. The former should be used sparingly because it will cause the band to bite or grab hold when forcibly applied and at times has been known to cut into and score a cast-iron drum. As a rule, planetary transmission bands should be handled in the same way as ordinary brake bands as to lining and relining, roughness of surface, lubrication, etc.

GEARS IN GENERAL

Since the whole subject of transmission concerns itself with gears, it will not be out of place to discuss the gears themselves and describe the many different kinds in use. Speaking broadly, the gears used may be classified according to the position of their axes, relative to one another. Thus we have axes parallel and in the same plane; parallel but not in the same plane; at right angles and in the same plane; at right angles and not in the same plane; at some other angle than a straight or a right angle and in the same plane; and the same, but not in one plane. These classes give us the forms of gear in common use, viz, spur gears, bevel gears, helical gears, herringbone gears, spiral gears, and worm gears.

Spur Gears. A spur gear is not only by far the most common kind of gear, but is also the easiest to describe, consisting, as it does, of a round flat disc with teeth cut in its circumference, that is, around the periphery of the disc, as shown in Fig. 32.

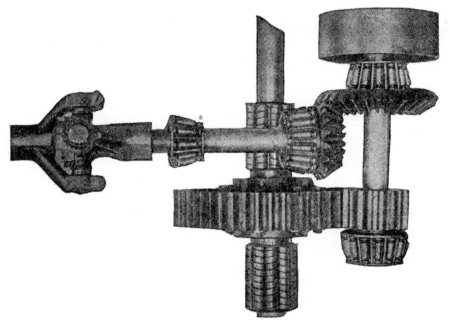

Fig. 32. Combination of Gears in the Autocar Final Drive

Bevel Gears. Bevel gears, in which the shafts are at right angles and in the same plane or in the same plane but not at right angles, are more difficult to cut and are therefore less used. They are now cut, like the spurs, in an automatic or nearly automatic machine, which requires little attention, but which does require more care than the spur-gear machine. Both spurs and bevels sometimes require a chamfered tooth edge. Spur gears as used in the Panhard, or clash-gear, transmission are always in need of it. This work was formerly done by hand, but now a special machine has been manufactured for this purpose.

GASOLINE AUTOMOBILES

There are no real restrictions against the use of the spur and bevel, either or both being used interchangeably. Very often they are used in combinations which appear peculiar, as the one shown in Fig. 32. This is the final drive and reduction gear of the Autocar commercial cars, made by the Autocar Company, Ardmore, Pennsylvania. In this gear, it will be noticed that the drive from the engine is through bevels to an intermediate shaft and that the final drive is by spur gears.

Helical and Herringbone Gears. In situations where quiet running is deemed necessary, the use of a helical gear frequently finds favor, since it accomplishes the desired result, although the cost of cutting is high. Of late, these gears have come into general use for camshaft drives and similar places. A pair of helical gears set so that the helices run in opposite directions forms a herringbone gear. This is even more quiet in its action than the single helix and possesses other virtues as well. One well-known firm has adopted it for camshaft driving gears and makes it as described to save cutting-cost, as the cost of cutting a true herringbone would be prohibitive. So a pair of helical gears of opposite direction are set back to back and riveted or otherwise fastened together, forming a herringbone gear at a low cost. Both of these may be used when the two shafts are parallel and in the same plane, but for all cases where the shafts are neither in the same plane nor parallel, some form of spiral gear must be made use of.

Spiral Gears. Spiral gears, as such, are not generally understood, but the variety known as the *worm* gear is very simple and easily understood and it has attained much popularity within the past few years. This popularity has been due, in part, to superior facilities for cutting correct worms and gears, but, in the main, to a superior knowledge of the principles upon which the worm works and of the things which spelled failure or success. Thus, one of the earliest experiments in this line laid down the law that the rubbing velocity should not exceed 300 feet per minute if success was desired or in rotary speed about 80 to 100 revolutions. For automobile use, this was out of the question, but later experimenters found that these results only attached to the forms of gear used by the early workers and did not apply to a strictly modern gear laid down on scientific principles.

The mistake made was in the pitch angle of the worm, which was formerly made small, nothing over 15 degrees being attempted. This was the item that was at fault and that caused this very useful and efficient mode of driving to fall into disuse. As soon as this fact was ascertained and larger pitch angles utilized, better results were obtained, until, with 20-degree angles, 700 feet per minute pitch-line velocity was attained, followed shortly by the use of even higher angles, resulting even more successfully. As the efficiency depends directly upon the pitch angle, these changes brought the efficiency of this form of gearing from the former despised 30, 40,

Fig. 33. Rear View of Timken Worm-Driven Rear Axle
Courtesy of Timken-Detroit Axle Company, Detroit, Michigan

and sometimes 50 per cent up to 87, 88, and even 90 per cent, thus putting it on a par with all but the very best of spur gears and above bevel gearing. In fact, in the light of modern knowledge of worm gears, it is possible to obtain from this form an efficiency of 93 per cent. In automobile work, the worm gear has been used mostly for steering gears and final drives. In the former, its irreversible quality is brought out, while in the latter, this quality must be made subordinate to a great reduction, which may be attained in a very small compact space. The majority of modern machines make use of worm gears.

Spiral Bevels. The spiral bevel is a new development, having been brought out in 1914 as a compromise between the worm and the straight bevel. As such, it is supposed to have practically all the advantages of both, except that it does not afford the great speed reduction that can be accomplished with a worm in the same space, being more like the bevel in this respect.

Among those using the spiral bevel are the Packard, Cadillac, Reo, Stearns-Knight, Velie, Kline, Apperson, Buick, Chalmers,

Chandler, Cole, Haynes, Hupmobile, Jackson, King, Locomobile, and many others. Fig. 33 shows a worm-drive rear axle: Fig. 34 shows the construction of the LaFayette spiral bevel and the differential gears. Fig. 35 shows the spiral gearing of the Cadillac passenger car.

Worm Gears. Progress in the application of worm gears for rear-axle use has been considerable in the last few years. In one

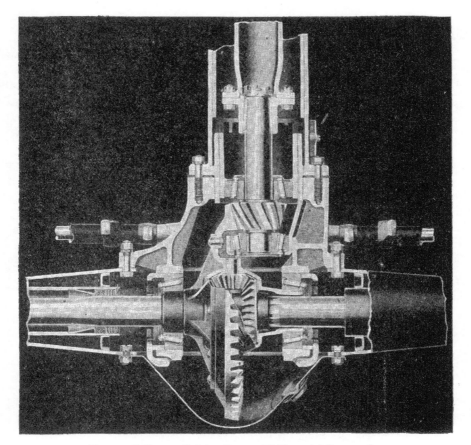

Fig. 34. LaFayette Driving Gear and Differential Construction

respect, at least, designers have found it an advantage. The top position for the worm was not much used at first, as it was thought impossible for it to receive sufficient lubrication. Consequently, it was placed in the bottom position, which cut down the road clearance considerably; in this position the clearance was less than with the ordinary bevel. With the proof that the worm could be lubricated in a satisfactory manner in the top position, the majority of gears are so placed, thus converting what was formerly a disadvantage into an advantage, for in the upper position the clearance

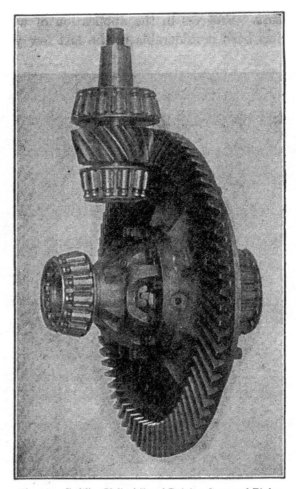

Fig. 35. Cadillac Helical-Bevel Driving Gear and Pinion

is greater than with bevel gears. This is shown quite clearly in Fig. 33, where it will be noted that the worm-gear housing in the center is actually higher than are the brake drums at either end of the axle, despite the fact that a truss rod passes beneath the center of the axle. For heavy trucks, especially, and for electric pleasure cars. the worm has proved an ideal drive. In these situations, there is the condition of high-engine or electric-motor speed, coupled with low-vehicle speed requirements, which necessitate a considerable reduction. As pointed out, the worm gives this in a small space.

Gear Pitch and Faces. The manufacturers of transmissions and of gears for them do not agree as to the best gears. Neither do they agree as to which gears are most quiet or most efficient. In general coarse-pitch stub-tooth gears are gaining faster than any other form. The 6–8 pitch is fairly general for gears of $\frac{3}{4}$-inch and $\frac{7}{8}$-inch face, and 4–5 pitch for wider gears. One manufacturer, Warner, considers the finer pitch gears and narrower faces as less likely to make noise, since they will not distort as much in hardening as wider gears. In this, other manufacturers agree, but there are some who claim to have had both quiet and noisy operation with both fine and coarse pitch. The tendency toward compactness has not increased transmission-gear faces any appreciable amount, nor has the increased use of better steels and better hardening processes lessened the size of the four noticeably.

Gear Troubles. There is not as much trouble with gears today as there was several years ago. This is due to better design, better materials, better processes, and better assembling on the part of manufacturers and to more skill in handling, caring for, and adjusting on the part of owners.

REBORING EQUIPMENT
Courtesy of Gisholt Machine Company

REAR AXLES—FINAL DRIVE

PART I

Units in the Final Drive. The transmission is generally located in the middle or forward end of the chassis. When this is the case, the final drive begins right at the rear end of the transmission. The units back of the transmission, then, would be a universal joint; a driving shaft; possibly another universal joint; the final gear reduction; rear-axle shafts and enclosure; the differential; the torque rod, or tube, or substitute for it; the wheels; the brakes; the tires; and other smaller units.

Even when the transmission is placed on the rear axle, this general layout is changed very little. In the case of a chain drive, which is still used on one pleasure car or perhaps two, on a number of small trucks, and on a large number of large trucks, this layout is changed considerably. In the large trucks, the transmission in perhaps 90 per cent of all cases would be in a unit with the jack-shaft, which means that for consideration in the final-drive group there would be only the driving shaft to the transmission; the joint or joints in it, if any; the chains and the method of adjusting them; the rear axle and wheels; the brakes; the differential, of necessity becoming a part of the transmission; and the jackshafts.

To make this clear and point out the various units, it will be noted in Fig. 1 that it is a unit power plant. Directly back of the transmission is the first universal joint, driving through the hollow propeller shaft to the rear axle, in front of which is the second universal joint. The rear-axle group includes the axle shafts, differential gears, final gear reduction, gear housing, and the wheels. The torque reaction of the drive is taken by the torque rod, marked in the drawing, which connects the rear axle to the under side of the stout frame cross-member in front of the axle.

Universal Joints. The purpose of taking up the universal joints is to show how the rear axle rises and falls or moves sidewise in either direction without making any difference in the transmission of power to the axle. When joints are used at other points, the

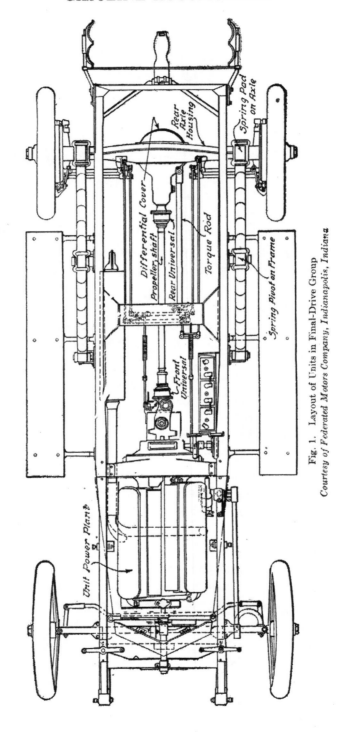

Fig. 1. Layout of Units in Final-Drive Group
Courtesy of Federated Motors Company, Indianapolis, Indiana

purpose is generally to take care of any lack of alignment, but here the purpose is to transmit power at an angle.

The transmission of power at an angle is effected by constructing the joint so that it can work at any angle. This is done by constructing the central member in the shape of a cross, with four projecting arms or pins, all in the same plane. The ends of the two shafts are made in the form of forks, or Y's, and are set at right angles to each other, that is, the forks are laid in planes which are at right angles. The fork on one shaft is fastened to a pair of diametrically opposite pins, while the fork on the other shaft is fastened to the other pair of diametrically opposite pins. Each shaft is able to turn on its pins about a line through the center of both. As

Fig. 2.   Thermoid-Hardy Universal Joint

these two lines are in planes which are at right angles to one another, but intersect at a common center, movement is possible in either plane, or by combination movements of both, in any direction.

Thermoid-Hardy Universal Joint. This universal joint, Fig. 2 is a coupling having the ends of the shafts permanently bolted to discs of flexible fabric in such manner that there are no metal-to-metal bearing surfaces. Metallic friction is thus eliminated, there being left only a small amount of friction caused by the distortion of the fabric discs. This type of universal joint has become very popular as it greatly simplifies the construction of the early type of universal joint and its manufacture is less costly.

In Fig. 3 is shown the method of cutting a disc. The individual layers of fabric are first placed in staggered form and then vulcanized together; this piece of fabric is then machine cut to disc form. The purpose in staggering these layers is to prevent them from being drawn out of shape; it also increases the tensile strength.

Fig. 3. How the Disc Is Cut

Tests have shown that these discs are capable of withstanding without damage to the fabric a driving torque that will twist a 2-inch 10-gage tubular propeller shaft. This form of joint is applicable to trucks as well as pleasure cars and is used extensively in their construction.

In Fig. 4 is shown the method of drilling the holes and the line of strain on the disc.

Metal Universal Joints. In this type of universal joints, the different parts are all metal and are made in several forms. (1) The bushing type—the bushings are held between two clamp rings,

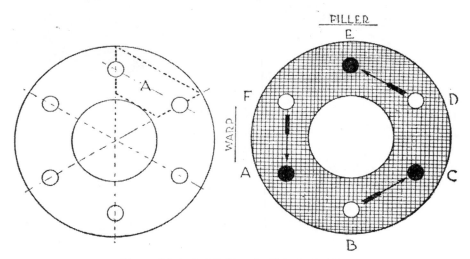

Fig. 4. Method of Drilling the Holes in the Disc

the bushings being free to move on the universal joint yoke pins, Fig. 5. (2) The center-block type—the center block is held in place by long pins which pass through the yokes, Fig. 6. (3) The circular-

54

yoke type—the rollers, which revolve on the yoke pins, are held in place by lock rings over all of which is placed an oil-tight cover, Fig. 7. The details of the universal joints shown in Fig. 5 are shown in Fig. 8. In the first type, great care should be taken in the fitting of the bushings for they must fit tight in the clamp rings but, at

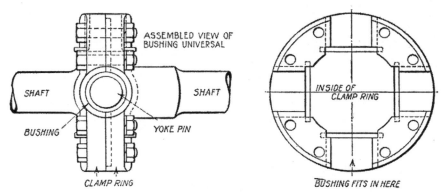

Fig. 5. Bushing-Type Universal Joint

the same time, must move freely on the pins. This type is not as easily lubricated as those that are enclosed in an oil-tight housing. In the second type, the thrust of the block comes on the inside of the yoke and causes wear which often means that all the parts must be renewed at the time of repairing. In the last type, the thrust

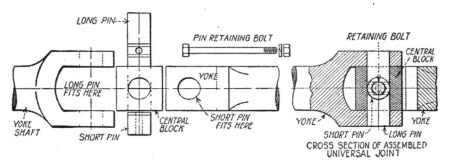

Fig. 6. Center-Block Universal Joint

is taken by the rollers and usually it is only necessary to renew the steel rollers at the time of overhauling.

Slip Joints. In many situations, a real universal joint is not needed, since the parts are not actually free to move in all directions; but what is needed is slight freedom up and down or sidewise combined with possible fore-and-aft movement. In such cases a slip is used, the name giving the idea of a joint which allows one shaft to slip, or slide, inside the other. The general construction of slip joints

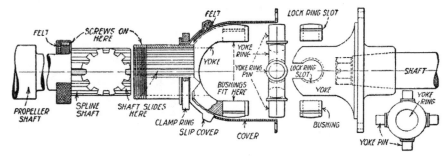

Fig. 7. Spicer or Circular-Yoke Type of Universal Joint

varies. Sometimes a round gear is fastened to the end of one shaft; this gear has a fairly large diameter and many teeth, with the teeth chamfered to an unusual extent—almost rounded, in fact. An internal gear of the same size and number of teeth with similarly rounded profiles is meshed with the hollow gear of the other shaft.

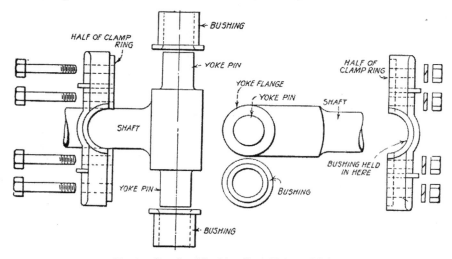

Fig. 8. Details of Bushing-Type Universal Joint

Both gears have unusually wide faces. This combination gives an action almost universal, and also allows lateral sliding of perhaps $\frac{1}{2}$ inch.

The second form of slip joint consists of a squared shaft and square enclosure.

Occasionally a square joint is constructed as simple and small as possible, in which case the housing is not split and the shaft end is not rounded. This gives a simple square which drives through a simple squared-out hole. In this case there is no universal action, but simply lateral or sliding freedom.

Other Flexible Joints. To get away from the complication of the universal joint and yet give practically the same results, many other forms have been produced. A very thin disc of tempered steel, with the two shafts bolted to the two opposite sides of it, has been used. The metal will bend and give enough to allow considerable angle of drive. Later forms of the same joint use leather in several thicknesses, the leather being bolted up to the two shafts in the same way. A joint of this kind, consisting of several layers of fabric which have been fastened together in laminations until a disc of fair thickness, say $\frac{1}{2}$ to $\frac{3}{4}$ inch, has been built up, is shown in

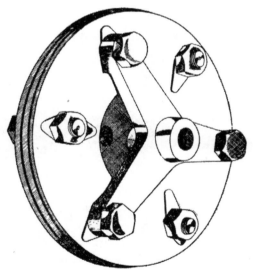

Fig. 9. Laminated Discs Forming Flexible
Shaft Coupling
Courtesy of Thermoid Rubber Company,
Fenton, New Jersey

Fig. 9. Then the leather is cut round and drilled for the bolts. In this form, six bolts is the preferred number, three for each shaft end; they are in a three-armed spider fastened to the end of each shaft, as the figure shows. These newer forms are usually convenient for the repair man, for they allow breaking into the main shaft by the simple removal of the three bolts (or two as the case may be). By taking out the bolts at each end of such a shaft, the shaft itself can be removed, leaving the other units in the chassis ready for immediate removal, according to the needs of the repair job.

Universal Joint Troubles. The disadvantage in the splined shaft and yoke is that the spline becomes worn and allows a great deal of play which can be plainly felt every time the engine is accelerated in the form of a jerk with a hammering noise. The only cure is to either install new yokes or shafts or to have the splines filled in by welding and recutting the splines. The propeller shaft is hollow as a rule with the opposite end of the splined part of the shaft welded and pinned into the shaft proper. If on acceleration there is a whistling noise, it is a good indication that the universal joints need lubrication. Rattling with back-lash and jerking when the car starts is an indication of play in the universal joints. The remedy is to install new universal joint parts.

Shaft Drive. In its usual form, shaft driving in an automobile involves simply a propeller shaft interposed between the rear axle and a revolving shaft in the car above the spring action. There is

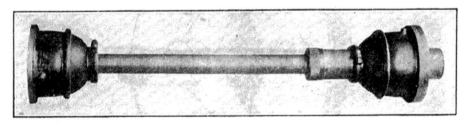

Fig. 10. Ordinary Driving Shaft of Solid Form with Two Universal Joints

some provision for taking the torque of the shaft and of the axle so that they shall maintain their proper relative positions.

In Fig. 10, a typical short driving shaft with its two universal joints is shown. This is such a shaft as would be used in the car shown in Fig. 1, except that the latter is a long wheel-base car with its transmission in a unit with the motor and clutch and thus, far forward. This combination necessitates a very long propeller shaft. The one shown is actually from a car having a short wheel base, with the transmission located amidships. This is a combination which calls for a fairly short propeller shaft.

The short shaft shown in Fig. 10 is a solid shaft. The modern tendency toward lighter weights is being worked out in the case of propeller shafts, and many are now made hollow. By making the diameter slightly larger and having a large central hole, unusually light weight is obtained with all the strength of the solid form. In addition, the larger diameter hollow shaft has more rigidity than the small diameter solid form, and in many of the modern cars without torque or radius rods, unusual rigidity of the driving shaft is necessary.

An objection to the shaft type of drive is that the reaction of the revolving shaft tends to tilt the whole car on its springs in a direction opposite to that in which the shaft is turning. In some cars, this is counteracted by the use of slightly heavier springs on

Fig. 11. Typical Roller Chain

one side. The advantages of the shaft drive are the complete enclosure of all working elements, with their consequent protection from dirt and the assurance of their proper lubrication.

As the axle rises and falls according to the conditions of the road, the distance between the universal joints is increased and decreased and there must be some provision made for this or else there will be a great strain thrown on the shafts and joints. In the flexible joint, the fabric gives and takes care of it; while in the metal joint, the yoke and the shaft are splined, which allows the shaft to slide in and out of the yoke, thereby increasing or decreasing the distance between the joints.

Double=Chain Drive. The use of double chains, by which the driving wheels of an automobile are driven from a countershaft across the frame of the machine, is a practice possessing a number of advantages. But because of the noise and quick wear with badly designed chain drives and the difficulties of completely enclosing the driving mechanism, chains are less popular than formerly. Nevertheless, the elimination of universal joints working through large angles and under heavy loads, the avoidance of heavy weights

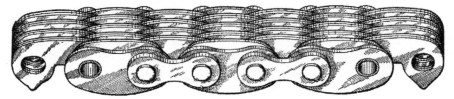

Fig. 12.   Typical Silent Chain

carried on rear axles without spring support, the lowering of the clearance by the differential housings, etc., are very real objections that the double chain avoids.

For trucking and other heavy service, chains are still commonly in use, and it is the belief of many that a better understanding of their merits and the means of securing these merits in positive and permanent form will result in their more general use.

A typical roller chain of the type most used for automobile drives is shown in Fig. 11.

Silent chains, of the types illustrated in Figs. 12 and 13, possess certain points of superiority over roller chains and are therefore com-

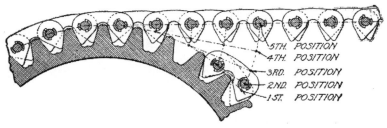

5TH. POSITION
4TH. POSITION
3RD. POSITION
2ND. POSITION
1ST. POSITION

Fig. 13.   Action of Silent Chain and Sprockets

ing increasingly into use for camshaft drives, in gear boxes, etc., and there is some possibility that they will find more extensive application to final drives.

The action of a silent chain is illustrated in Fig. 13, in which it is seen that as the chain links enter the sprocket teeth the chain teeth at the same time close together and settle in the sprocket with a wedging action that causes them to be absolutely tight, but without any more binding than there is backlash.

To keep silent chains from coming off sidewise from the sprockets over which they run, it is customary to make the side links of deeper section than the center links, as is illustrated in Fig. 12. Another successful scheme is grooving the sprocket to receive a row of special center links in the chain, which are made deeper than the standard links.

At the present time there is no American car of any note using the chain drive. Chains are chiefly being used on trucks, more especially on those that are used for heavy duty work such as the Mack and Sauer.

Torque Bar and Its Function. It is a well-known fact that action and reaction are equal and opposite in direction, so that if a gear is turned forcibly in one direction, say clockwise, there is a reaction in the opposite direction, or counter-clockwise. This is the simple basic reason for a torque bar, or torque rod, on an automobile. It is needed with any form of final drive, but it takes differ-

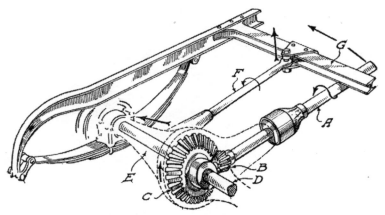

Fig. 14. Diagram to Show What Torque Is and Why Torque Rods are Necessary

ent forms, according to the type of gear used. Fig. 14 shows the rear end of a typical pleasure-car chassis. The engine is rotating clockwise, and so is the driving shaft A, as shown by the arrow. The shaft turns the pinion B in a clockwise direction, which rotates the large bevel C so that its top turns toward the front of the car. The bevel C turns the rear axle D and the rear wheels (not shown) in the same direction; so the car moves forward.

In addition to the gear C and shaft D turning easily in the axle housing E, there is an equal and opposite reaction which tends to keep them stationary, while the bevel pinion B and driving shaft A tend to rotate around the rear axle as a center in a counter-clockwise direction, as shown by the diagram. If the rear axle were held firmly so it could not rotate and there was nothing to restrain the bevel pinion and shaft, this could easily happen. However, a means is provided to oppose this action and prevent it from happening. Since the turning force which makes the shaft rotate is called the torque, this rod, bar, or tube, whatever its form, is called the torque member.

In the sketch, the torque member is marked F and is attached to the frame cross-member G, between a pair of springs, so as to

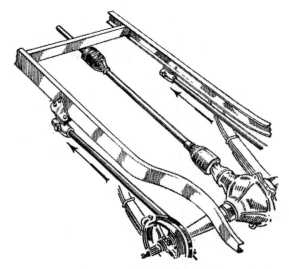

Fig. 15. Diagram to Explain Driving Reactions
Using Radius Rods

cushion the shocks of sudden car or shaft movements. The force on this is the force which tends to rotate the driving shaft and pinion counter-clockwise, so that it works upward, as shown by the arrow. The frame prevents this and absorbs the force.

Driving Reaction. As has been stated, the power, or torque, of the motor is used to rotate the rear wheels. These stick to the pavement or road surface, so the car is really pushed forward. Since it is this pushing action which really moves the car forward, it is very interesting to note how this push is transmitted from the wheels and rear axle to which they are attached to the frame which carries the body and passenger load.

GASOLINE AUTOMOBILES

The transmission of the drive to the body is accomplished in one of three ways. The first form was the so-called radius, or distance, rod, which the shaft-driven car inherited from the chain-driven form. In the chain drive, these rods were a necessity and served a double purpose; they kept the driving and driven sprockets the proper "distance" apart for correct chain driving (hence their name "distance" rods), and they also transmitted the drive back to the frame. On the shaft-driven car, the distance function is not needed, so they are called radius rods. As shown in Fig. 15, they transmit the drive forward to the frame, thus propelling the car in

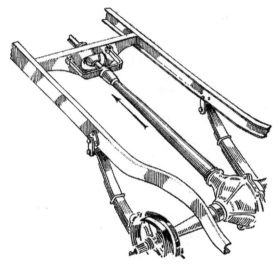

Fig. 16.   Layout of Driving Reaction Using Torque
Tube around Shaft

the direction of the arrow, and they also keep the rear axle in its correct position.

In lightening and simplifying the shaft-driven car, designers figured that three members for the torque and driving reactions were too many; so a design was worked out in which all three were combined into one, which is a form of tube surrounding the shaft. This made the member light but strong, and simplified the whole rear end. As shown in Fig. 16, the tube has forked ends at the front, which are connected to the frame cross-member in such a way as to absorb the torque reaction and also to transmit the drive. The method has the further advantage of needing but one universal joint, and that at the front end. Furthermore, it gives a correct radius of rise and fall for the rear axle, since the center of the combined torque and drive member is also the center of the universal joint in the driving shaft. In the form shown in Fig. 14 (radius rods not shown), the two different centers will be noted, the torque rod giving a greater radius than the shaft. Similarly, in Fig. 15 (where the torque rod is omitted for clearness), the rods give a longer radius of rear-axle movement than the shaft, which has a joint close to the axle.

It will be noted that both these methods allow complete freedom of the rear springs, which may be of any form, and shackled at the front end if desired by the designer. In its newest and simplest form, the so-called Hotchkiss, or spring, drive has both the radius and torque rods omitted, the springs being forced to transmit both

Fig. 17. Arrangement of Driving Reaction When Hotchkiss Drive is Used

forces, as shown in Fig. 17. In this case, the forward end of the rear spring must act as a rod, or lever, instead of as a spring, and must be fairly straight and stiff without a shackle, but firmly pivoted on the frame. In addition, the shaft must have two universal joints, as shown.

This last form is increasing rapidly and at the expense of the other two. On smaller lighter cars it is gradually replacing all other forms. It has the advantages of minimum weight and fewer parts, and applies the driving force in a direct line to the frame, the same as the two radius rods do. On the other hand, it makes the springs serve a triple purpose, the demands on these for torque and drive transmission and absorption being such that the spring flexibility must be negligible, which makes the car ride hard. In addition, making the springs handle the three widely different actions

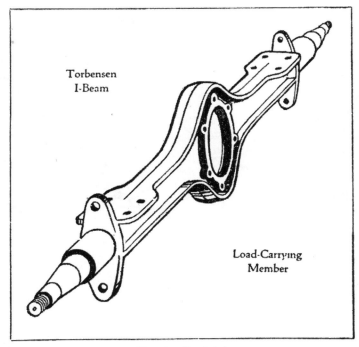

Torbensen
I-Beam

Load-Carrying
Member

Fig. 18.   Dead Axle

puts additional stresses upon them, so that they are more likely to break. On the medium size and larger heavier cars, this construction is not gaining so rapidly.

REAR AXLES

Classification. Rear axles may be divided into two classes as used on the pleasure car and truck—the dead axle and the live axle. The dead axle is one that supports the weight of the car or truck only and has no work to do with regard to driving except to act as a support to the parts that do the driving. In a truck with a chain drive, the dead axle supports the weight and the wheels revolve on the end of the dead axle, Fig. 18. The live axle takes the weight through its housing and the axles do the driving of the wheels but the two are a complete and self-contained unit. The live axle is often used in conjunction with the dead axle as shown in Fig. 19.

Live Axles. There are three distinct types of axles used under this classification—(1) the semi-floating form, carrying the drive and all of the load on the axle, the axle shafts not being removable without removing the wheels; (2) the three-quarter floating form, carrying the drive and a small part of the load, the latter being

Fig. 19. Rear Construction Embodying Dropped Type of Rear Axle

divided between the shaft and its housing, but with the shafts removable; seven-eighths floating form, carrying the drive but not the load, the arrangement of bearings to take the load being such that the wheel hubs do not rest wholly and solely upon the axle-casing end; and (3) the full floating form, in which the shaft does nothing but drive, and is removable at will without disturbing the wheel and wheel weight resting on the axle-casing end, which is prolonged for this purpose.

With the full floating form, any accident to the wheel, in which it was struck from the side, also damaged the casing, or tube, end. The result of this in nine cases out of ten was to make the removal of the wheel impossible, because the tube end, which projected through, was bent over. Moreover, repairing in such a situation called for a new axle casing—a very expensive proposition. Consequently, the seven-eighths floating form was developed to present all the advantages of the full floating form, with this serious drawback eliminated by a rearrangement of the parts which did not

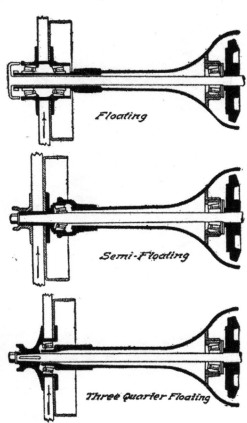

Fig. 20. Arrangement of Axle Bearings and Housing in Three Principal Forms of Rear Axle

necessitate prolonging the axle through the wheel hubs. However, it did not gain as rapidly as the other floating forms, and now is almost out of use.

The three diagrams in Fig. 20 explain the types as well as words can. At the top is shown the full floating axle, the best but most expensive form. In the middle, the semi-floating axle, which makes the axle shaft do all the work—carrying load as well as transmitting power—is shown. At the bottom is the three-quarter floating form, which is really a combination of the other two forms and possesses a maximum of advantages with a minimum cost. The car weight is carried on the tubing, while the shaft drives and carries a portion of the side stresses to which the wheels are subjected, the quantity depending upon the construction of the bearings.

Fig. 21 shows a full floating axle, with the ends of the driving shafts projecting beyond the housing and carrying five jaws which mesh with five similar ones in the wheel hubs and thus drive the wheels. Unless the jaw end is welded on to the shaft, this makes a very expensive axle despite its many good points. Fig. 22 shows

Fig. 21.   Example of Full Floating Type of Axle

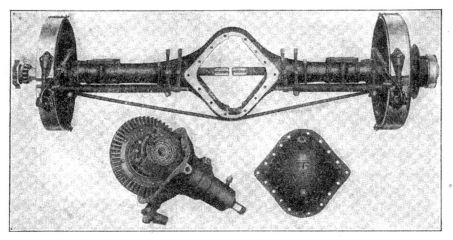

Fig. 22.   Timken Full Floating Rear Axle, Showing Differential Removed
Courtesy of Timken-Detroit Axle Company, Detroit, Michigan

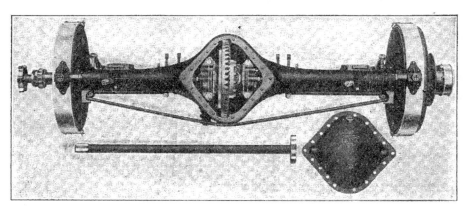

Fig. 23.   Timken Full Floating Rear Axle with Spiral Bevel Gears

the rear construction of a car with full floating axle, with the brace below it for the purpose of strengthening the whole construction.

The large diameter brake drums, shown close to the wheels, are made of pressed steel and are united to the axle tubing, which is also united to the differential housing, so that the whole forms one large and continuous piece, except where the differential unit bolts on one side and the cover on the other. Note that the shaft has the driving clutches machined as an integral part, and that removing the two shafts for a few inches makes it possible to unbolt and remove the entire differential unit. For the sake of comparison, Fig. 23 shows an axle which differs from Fig. 22 only in having

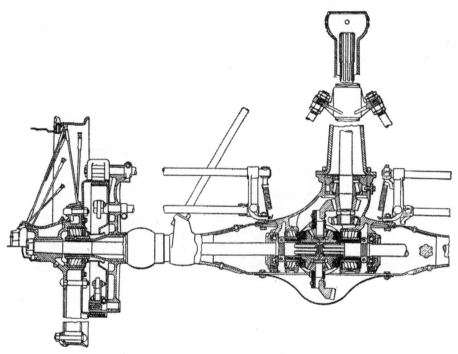

Fig. 24. Partial Section through Rear Axle of Case Car, Showing Construction
Courtesy of J. I. Case T. M. Company, Racine, Wisconsin

spiral bevels substituted for the ordinary straight-tooth bevels. In Fig. 23, the differential unit is removable in the same manner as in Fig. 22. One of the axle shafts, with its integral driving clutches, and the differential cover are shown below. Note the two plugs in the cover; the upper one is for filling the case with lubricant, while the lower plug acts as a level indicator. When it is opened, heavy oil or oil and grease combinations are put in the filling plug above until the lubricant begins to flow out of the lower opening.

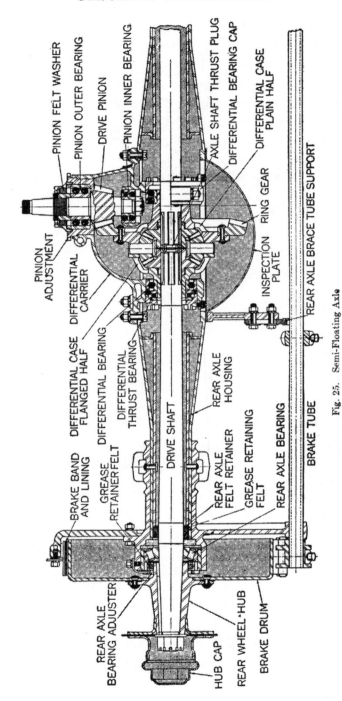

Fig. 25. Semi-Floating Axle

A rear axle of the three-quarter floating type is shown in Fig. 24. Note the enclosure of the driving shaft and the splines at its forward end for the universal joint, also the housing for the joint forming the torque member. The small roller bearing for the spigot end of the driving shaft beyond the bevel pinion is unusual; so are the diagonal distance rods, the spherical seat for the springs, the combination of drawn-steel tubes, steel castings, and pressed-steel cover for the axle housing. The wire wheel and its method of attachment will be seen, also the double set of brakes, internal and external

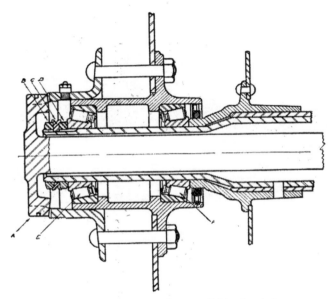

Fig. 26. Wheel Attachment Full Floating Axle

on the same drum, with operating shafts for both supported from the central part and ends of the axle housing.

The semi-floating type of axle is shown in Fig. 25. The difference in its construction may be seen by comparing it with Figs. 23 and 24. The construction of all semi-floating axles is not alike. In the axle shown, the meshing of the driving pinion and the ring gear adjustment can be made when the inspection plate has been removed. In others, such as in the Ford and Chevrolet, the adjustment is fixed. If there is an adjustment to be made, the whole axle must be removed and taken to pieces.

Wheel Attachments in Full, Three=Quarter, and Semi=Floating Types. Apart from the difference in design between the three types of axles, there is a difference in the attachments of the wheels to the axles and the manner in which the wheel is driven by the axle. In the full floating type there are slots cut in the hub of the wheel, and on the axle is a flange in which is cut or machined projections or dogs which fit tightly into the slots cut into the hub. In this manner the wheel is driven, the shaft being held in place by the hub cap and a spring. Fig. 26 shows the construction of the outer end

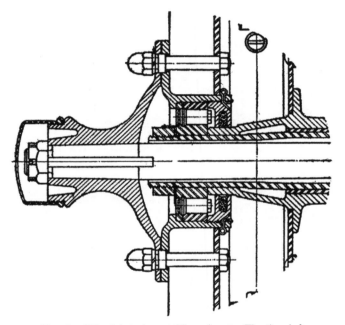

Fig. 27. Wheel Attachment Three-Quarter Floating Axle

of the axle with the bearings in place and the axle can be removed without disturbing the wheel. The hub cap is already removed in the view shown.

In the three-quarter floating type, Fig. 27, the axle is attached to the hub flange by a taper on the axle and a nut. The flange is held to the wheel hub by bolts upon which are acorn nuts. To remove the axle it is only necessary to remove the acorn nuts, which may be done without disturbing the wheel. The advantage of the two foregoing installations is that adjustments can be made to the wheel bearings at the end of the axle without disturbing the wheels.

In the semi-floating type, Fig. 28, the wheel is attached to the axle by taper, key, and nut, as in the three-quarter type, but before any adjustment can be made to the bearings, the wheels must be taken off of the axle. In this type of drive, the axle and wheels must be removed before the differential unit can be removed, while in the three-quarter and full floating, the axle can be removed and the differential taken out and the car allowed to stand on the wheels,

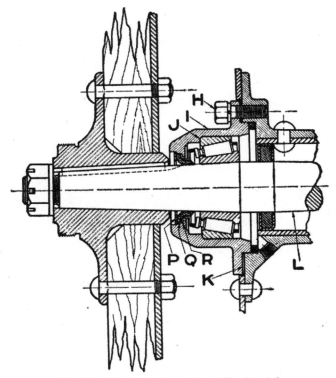

Fig. 28. Wheel Attachment Semi-Floating Axle

which is safer and much easier for the repair man. It is quite an asset to the repair man to be able to recognize on sight the three types of axles. If he is able to do this, he will know how to start to do any repair work without waste of time. The full floating is recognized by the shape of the hub after the hub cap is removed and by the slots. The three-quarter is recognized by the nut on the outside of the wheel hub. The semi-floating is recognized by the absence of the nuts on the outside or by the shape of the hub after the cap has been removed, and by the absence of the slots.

Dropped Rear Axle of Full Floating Type. The dropped type of axle is not much used at present for cars of the shaft-driven type, the dropped part of the axle bed being used to hold the rearward-placed transmission. Fig. 19 shows a former American type, in which the weight of the car as well as the weight of the load is carried on the I-section drop-forged rear axle, while the drive is transmitted from the transmission by the usual shafts, which carry no load. The cut shows the complete assembly above and the dropped axle below. The round ends of the I-beam axle are hollow, carrying the driving shaft through the central hole and the wheels on bearings which fit over the outside. The wheels will revolve on the bearings, even if the inner shafts and transmission be removed from the chassis.

Despite its manifold advantages, the expense of constructing an axle of this type—it is practically the same as that of two ordi-

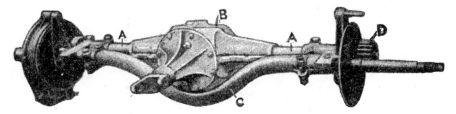

Fig. 29. Rear Axle, Showing Wheels Driven by Spur Gears

nary axles, making the total cost double that of any other form—has worked to prevent its general use. But it is being used in trucks today and found to give excellent service.

The dropped type of axles are neither all shaft-driven nor all chain driven. Fig. 29 shows one that is of the spur-gear driven type. The dropped axle bed C is of tubular form, and the differential case is dropped down on and slightly back of the rear axle, as at B. From this case, two shafts AA extend out to the sides, driving the wheels through the medium of the spur gear D, which meshes with internal gears within the wheel hubs (not shown). This type of rear axle and drive is used on a number of the Fifth Avenue stages in New York City.

Internal=Gear Drive for Trucks. The spur-gear driven type just described is gaining rapidly for motor truck use, because it has a number of important advantages. Besides carrying the heavy load on a member able to withstand any amount of overload, it

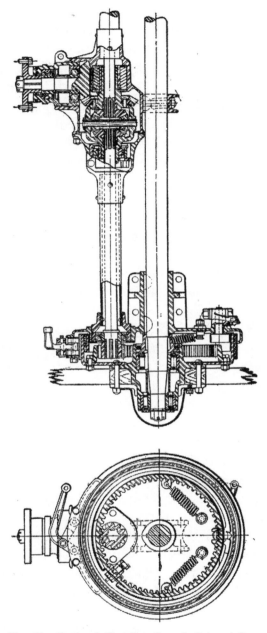

Fig. 30. Sectional Drawing through Internal-Gear
Drive Axle of Three-Quarter Ton Capacity
Courtesy of Russell Motor Axle Company,
Detroit, Michigan

materially lightens the power-transmitting portion of the axle, which is enclosed and therefore quiet. It is simple and inexpensive to construct and repair. Fig. 30 shows a section through one of these axles, which is used on a very light truck of $\frac{3}{4}$-ton capacity. It will be noted that the load-carrying axle is behind the power-transmitting shafts, consequently the former is straight. In Fig. 29, the load-carrying axle is in front and consequently must be bent down at the center. This bend is a source of weakness.

Rear=Axle Housings. Rear-axle housings are usually of pressed steel, although castings play a very important part and are sometimes used alone and sometimes in combination with other castings or in combination with pressed steel. Aluminum, although not a dependable metal, is used quite a good deal for the purpose of saving weight, as excess weight upon the rear axle is anything but desirable. In one unusual but effective combination, the axle housing consists of two malleable-iron castings joined together by means of bolts at the centers, the brake drums being cast as a part of the tubes. While not usual, this is safe practice, for malleable iron is tough and will not break or splinter. It seldom is the case, however, that the axle casing is reduced to as few parts as are shown here.

Welding Resorted To. Where the differential housing or brake drums are of malleable iron, cast steel, or even of pressed steel, and it is desired to unite them with the steel tubing forming the main part of the shaft housing, welding is now universally used. Formerly, it was good practice to make the casing a drive fit on the tube, riveting it in place, or else soldering it in place, making doubly sure by using rivets. Now, however, welding is resorted to, either the oxy-acetylene, electric, or some other process being used.

In the axles shown in Figs. 22 and 23, it will be noted that the axle shell is of pressed steel, to which the spring seats are bolted, the remainder of the construction being formed by drawing. In Fig. 21, however, the construction is such as to necessitate making the two halves longitudinally and then bolting or spot-welding them together. Being machined after they are fastened together, it makes as accurate a construction as the one-piece jobs, Figs. 22 and 23.

REAR AXLES—FINAL DRIVE

PART II

Types of Final Drive. There are practically three types of final drive gearing. Two of these are actually gears and the other is a worm drive. The two types of gears actually used in the gear type are the straight bevel gear and the helical or spiral gear. Fig. 31 shows the Ford final drive, which is a straight bevel drive. Fig. 32 shows a ring gear and a drive pinion of the spiral gear type. Fig. 33 shows a worm installation. The helical or spiral gear type is

Fig. 31. Ford Final Drive, Differential, and Axles

the most popular and has several distinct advantages over the straight bevel type, the chief of which is its silent operation at all speeds. The minimum number of teeth in the helical gear can be less than that of the straight bevel gear. The fact that as the number of teeth in the gear is reduced so is the uniformity of motion, which is the controlling factor in the number of teeth used in the gear; and as the number of teeth is reduced, the noise in operation is increased. The helical gear is naturally far more silent and therefore the number of teeth is fewer and consequently the size of the gear may be smaller. The worm gear drive is more suitable to trucks and slow moving vehicles than to the fast moving pleasure car. The worm gear has the advantage that a large gear reduction can be obtained by its use when compared with the bevel type of gears.

Effect of Differentials on Rear Axles. A differential gear, sometimes called a balance, or compensating, gear, is a mechanism which allows one wheel to travel faster than the other and which at the same time gives a positive drive from the engine. This device is a

Fig. 32. Spiral Bevel Gears—a New Noiseless Type
for Rear Axles
*Courtesy of Timken-Detroit Axle Company,
Detroit, Michigan*

necessity in order to allow the car to go around a curve properly, for in doing so the outer wheel must travel a greater distance than the inner one during the same interval of time.

There are two forms of differential, the bevel type and the all-spur type, the latter differing from the former only in the use of spur gears instead of bevel gears. The principle used in both is that a set of gears are so held together that when a resistance comes upon one part of the train of gears the whole train will stop revolving around on a stationary axis and revolve around another gear as an axis, the first gear, in the meantime, standing stationary, or practically so, according to the amount of the resistance encountered. In the bevel type, a pair of bevels are set horizontally. Between the bevels is a spider with three or four arms, with a small bevel on the end of each. These small bevels mesh with the larger bevels at the sides and ordinarily stand still, rotating around on the arm of the spider as an axis by virtue of the continued rotation of the two

Fig. 33.   Worm and Gear for Rear Axle, Showing Upper Position of Worm
Courtesy of Timken-Detroit Axle Company, Detroit, Michigan

side gears in opposite directions. When one wheel meets greater obstructions on the road than the other, thus holding it back, the shaft which drives that wheel lags behind the shaft driving the other wheel and thus holds back the horizontal gear attached to the shaft. This retarding movement allows the other horizontal gear more freedom to rotate. The result is that the spider carrying the smaller bevels rotates around on its axis, thus imparting to the free gear attached to the free wheel an additional motion, and to the free wheel a doubled speed, while the retarded wheel has a lessened speed. This takes the car around the corner without breaking the rear axle, as would be the case without some such contrivance. The description of the bevel differential action applies equally well to the spur type, except that all gears are spurs.

The dividing of the rear axle is, of course, done to make a place for the differential gear to work, and much time and thought have been given to this subject in an endeavor to work out a substitute which would permit the differential action and still allow the strength-

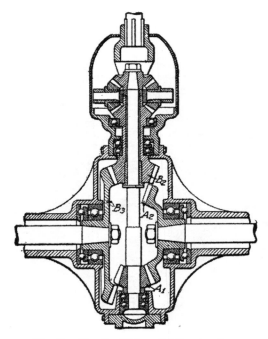

Fig. 34. Peculiar Differential Construction

ening of the rear axle. Fig. 34 shows one solution of the problem, which has been worked out in such a way that the differential is moved forward into the driving shaft. The rear axle shafts are thus greatly strengthened, the designer being unhampered by the presence of the differential in the rear axle. In this design, one side gear of the bevel-gear differential is carried upon a shaft, and the other upon a tube around the shaft. Then, at the rear axle, two sets of bevel gears $B_2 B_3$ and $A_1 A_2$ are used, A_1 being driven by the main shaft, and driving the right-hand shaft through the gear A_2; while the other B_2 is driven by the tube, and drives the left-hand shaft through the gear B_3. In this case the axle shafts are made much larger than in the ordinary case, while the differential action is just the same.

Improved Forms of Differential. Lately, much work has been done upon differentials to cause them to act as differentials should. The present form of differential acts according to the amount of resistance offered, but should act according to the distance traveled. When no resistance is offered, all the power is transmitted to that wheel, leaving the other stationary. This is just the opposite of the desired effect. If a differential were constructed to work for distance only, then, in the case of a wheel on ice or other slippery surface which offered little or no resistance, both wheels would still be driven equally, and the power transmitted to the one not on the ice would pull the vehicle over it.

One way in which the differential action might be corrected is by the use of helical gears and pinions instead of the usual bevel or

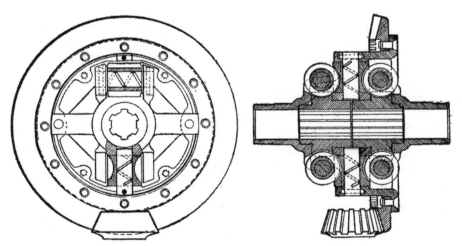

Fig. 35. The M & S Helical Gear Differential in Sections
Courtesy of Brown-Lipe-Chapin Company, Syracuse, New York

spur gears. In the M & S forms, this construction is used, Fig. 35, showing the form constructed by Brown-Lipe-Chapin. In this form, each axle shaft carries a helical gear, and the differential spider carries two helical pinions with radial axes and four additional pinions, each of which meshes with one of the radial pinions and one of the gears on the axle shafts. On a turn, the outer wheel tends to run ahead of the inner and thus causes the nest of helical gears to revolve. All gears and pinions have a right-hand 45-degree tooth, so that one wheel may revolve the housing if the other is locked or held, but it is impossible to turn the free road wheel by pulling on the housing. The principle is the same as a worm steering gear in which the turning of the hand wheel may be transmitted to the front wheels, but the gear cannot be operated from the wheel end, because

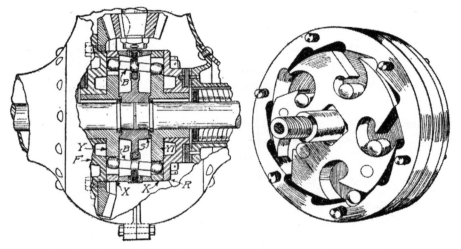

Fig. 36. Sketches Showing Construction and Operation of Gearless Differential

the worm is irreversible. This differential is used to advantage to prevent spinning on slippery ground and also to eliminate the skidding which the ordinary differential gives.

Another somewhat similar device has but two pairs of helical pinions in addition to the two helical gears on the shafts, the axes of each pair being set at an angle to the others. Thus, each helical gear and its pinion form an irreversible gear combination, so that movement cannot be transmitted through either in the reverse direction. This form fulfills the same conditions as the Brown-Lipe-Chapin M & S form, as the construction is such that no motion can be transmitted from the differential spider or housing to one of the wheels alone.

The above principle is back of the gearless form, shown in Fig. 36, in which the result is achieved through ratchets instead of helical gears, the lack of gears giving it its name. In this form there are two ratchets *Y* and *Y1*, which are keyed to the two axle shafts and free to rotate independent of the housing. The round members marked *B* are the interlocking pawls; the upper one is in a tooth of the right-hand ratchet at the right and is driven by the contact face of the driving sector *X* at the left. Thus, the right-hand ratchet is being driven positively forward. The lower pawl is engaged at the other end; so the left-hand ratchet is also being driven positively forward. On a turn, one wheel revolves faster than the other, say the right, and causes the right-hand ratchet to move faster than the differential housing, which can only go as fast as the other, or slow-moving, wheel. Then, the right-hand ratchet pushes the end of its pawl out of the tooth and gives it a free movement forward. As soon as the wheels revolve at equal speeds, the spring pushes it back. In the figure, the right-hand portion shows the original form in perspective.

Possible Elimination of Differential. The whole modern tendency is toward differential elimination. In the cyclecars and small cars brought out in recent years, designers have been forced to get along without it because of the demand for simplicity, light weight, and low price. This effect has been obtained by the use of a pair of driving belts, letting one slip more than the other; by the use of friction transmissions; by simply dividing the rear axle and letting one side lag when there was resistance; by not dividing it and letting one wheel drag; and in other ways. The evident success of these small vehicles without a differential or any real substitute for one has set designers to thinking about this subject again, and some big cars without a differential, or with a more simple and less expensive substitute for it, may appear in the near future.

Rear=Wheel Bearings. The bearings used on rear axles differ very little from those used on front axles. All forms are used—plain bearings, ball bearings, ball thrusts, roller bearings in both cylindrical and tapered types, and all combinations of these. Thus, Figs. 34 and 37 show the exclusive and liberal use of ball bearings, while Fig. 24, Part I, shows all rollers of two kinds and ball bearings for thrust bearings only. The two kinds of roller bearings are the tapered roller and the flexible roller. In Fig. 22, Part I, it will be noted that balls are used with two kinds of rollers, straight solid rollers in the wheels and flexible rollers in the differential case. Figs.

27, and 28, Part I, show the exclusive use of the tapered roller type, a construction which is gaining ground very rapidly, although, formerly, ball bearings were most widely used. The materials employed are similar to those used for front axles, which have been previously described. Cases are made of all kinds of steel and iron— pressed, drawn, cast, etc.—not to speak of crucible steel, malleable

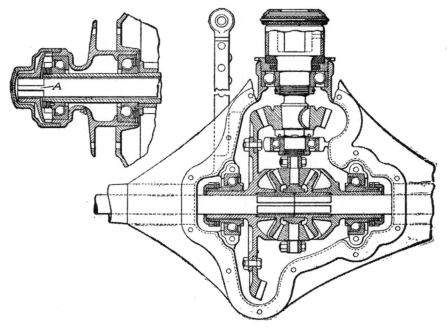

Fig. 37. Typical Ball-Bearing Differential

iron, manganese bronze, phosphor bronze, aluminum, aluminum alloys, and many combinations of these materials in twos and threes.

Rear=Axle Lubrication. Rear-axle lubrication is generally automatic in so far as the central bevel or other gears and the differential housing are concerned. The housing usually has a form of filling plug, or standpipe, which is used to fill the case with a form of heavy grease every 5000 miles, or once each season. The case is generally arranged so the filling plug works through and lubricates the outer bearings on the axle shafts as well, with suitable provision against this reaching the brake drums or other brake parts. The wheel bearings either are cared for in this way or have a central space which is filled with heavy grease once a season, being self-lubricating from then on. Such other rear axle parts as need occasional lubrication, as torque-rod pivots, brake-band supports, brake-operating shafts, etc., are generally provided with external grease cups, which are given a turn once a week on the average. It is highly important that the braking system be as well lubricated as the lubricating means provided will allow.

REAR=AXLE TROUBLES AND REPAIRS

Jacking=Up Troubles. Much rear-axle work—practically all, in fact—calls for the use of the jack. True, the full floating type of axle can have its shaft removed without jacking, but, aside from differential removal, there is little rear-axle trouble in which it is necessary to remove the shaft alone. In almost all cases, the axle must be jacked up. Many axles have a truss rod under the center, and this is in the way when jacking; however, this can be overcome. Make from heavy bar iron a U-shaped piece like that shown on

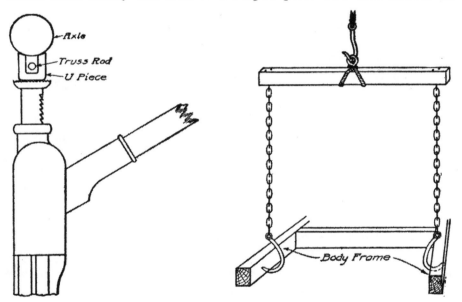

Fig. 38. Simple Arrangement for Avoiding Rear-Axle Truss Rod

Fig. 39. Simple Automobile Frame Hoist

top of the jack in Fig. 38, making the width of the slot just enough to admit the truss rod. The height, too, should be as little as will give contact with the under side of the axle housing.

Substitute for Jack. A good substitute for a jack is a form of hoist, Fig. 39, which will pick up the whole rear end of the car at once. This not only saves time and work, but holds the car level, while jacking one wheel does not. Moreover, with a rig of this kind, the car can be easily lifted so high that work underneath it may be easily done. The usual hoisting blocks are very expensive, but the above hoist can be easily made by the ingenious repair man. This one was made from an old whiffletree with a chain attached at each end. For the lower ends of the chains, a pair of hooks are made sufficiently large to hook under and around the biggest frame to be handled. With the center of the whiffletree fastened to the hook of a block and tackle, the hoist is complete. By slinging the hooks under the side members of the frame at the rear, it is an easy matter to quickly lift that end of the chassis any distance desired.

Workstand Equipment. Next to raising the rear axle, the most important thing is to support it in its elevated position. To leave it on jacks is not satisfactory, for they will not raise the frame high enough, and, furthermore, they are shaky and may easily let the whole rear end fall over, doing considerable damage. With the overhead hoist, the chains or ropes are in the way; so a stand is both

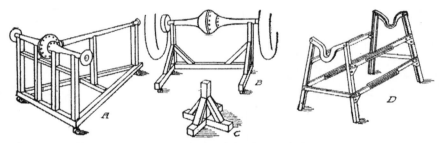

Fig. 40. Types of Handy Stands for Rear-Axle Repair Work

a necessity and a convenience. In Fig. 40, several types of stands are shown. *A* is essentially a workstand, intended to hold the axle and part of the propeller shaft while repair work is being done thereon. It consists of a floor unit, or base, built in the form of an *A*, with six uprights let into it, preferably mortised and tenoned for greater strength and stiffness. Then, the four rear uprights are joined together for additional stiffness and rigidity. If casters are added on the ends, the stand can be more conveniently handled around the shop.

The forms *B* are for more temporary work and consequently need not be so well or so elaborately made. The little stand *C* is a very handy type for all-around work. Stands of this kind with the top surface grooved for the axle are excellent to place under cars which have been put in storage for the winter.

The stand D is, like A, a workstand pure and simple. In this, however, the dropped-end members allow supporting the axle at those points, while the elimination of central supports gives plenty of room for truss rods. This type of stand would preferably be made from metal, pressed steel or small angle irons being very good. Every repair shop should have a considerable number and variety of stands, made as the work demands them, to fit this particular class of work.

Truss Rods. Truss rods hold the wheels in their correct vertical relation to the road surface and to one another. If, through wear or excessive loading, the axle sags so that the wheels tip in at the top, presenting a knock-kneed appearance, the truss rods must be tightened up. Usually, they are made with a turnbuckle set near one end, a locknut on each side preventing movement. The turnbuckle is threaded internally with a right-hand thread on one end and a left-hand thread on the other, so that a movement of the turnbuckle draws the two ends in toward one another, shortens the length of the rod, and thus pulls the lower parts of the wheels toward one another, correcting the tipping at the top.

To adjust a sagging axle, loosen both locknuts, remembering that one is right-handed and the other left. Then, with the wheels jacked clear of the ground, tighten the turnbuckle. A long square should be procured or made so that the wheel inclinations may be measured before and after. Placing the square on the ground or floor, which should be selected so as to be perfectly level, the turnbuckle should be moved until the tops appear to lean outward about $\frac{1}{2}$ inch—some makers advise more.

It should be borne in mind that even if the wheels and axle do not show the need of truss rod adjustment, if this rod be loose, it will become very noisy and rattle a great deal, as the rear axle sustains a great amount of jouncing. Moreover, this noise and rattle, if not taken up, will cause wear, which cannot be taken up.

Disassembling Rear Construction. In disassembling the rear construction for purposes of adjustment or repair, the repair man should be careful to mark all parts. Those parts which have been running together for several thousand miles act better and with less friction than would those which have never run together, despite the fact that the duplicate parts are supposed to be alike and interchangeable. It is therefore suggested that separate boxes be provided for the parts taken from the two ends or sides. The method of disassembling is about as follows: Jack up the axle, replacing the jack with small horses or blocks of wood if possible. Take off the hub caps, then free the wheels and take them off. Disconnect the brake-operating rods and levers and remove them from the car, marking them carefully. Spread the brake shoes apart, loosen the springs at one side, take out the springs, and then loosen and take out the brake shoes themselves. Remove the brake operating shaft with the cams; then disconnect the spring bolts and jack up the chassis, using the spring for a support. Disconnect all torsion or radius rods and take off the grease boot around the universal joint in the driving shaft. Open this joint and disconnect the shaft. Take this off and if the spring bolts have been removed, the rear axle will be free. Pull it out from under the chassis, and, if desired, further disassembling may be done more easily with the member clamped in a vice or laid on a bench.

Assembling. In assembling, almost the reverse of this process is followed, the parts going together in the opposite manner from that in which they were taken down.

Noisy Bevel Gears. Bevel gears make a noise because they are poorly cut, because they are not set correctly with relation to each other, or because the teeth have become cut, or chipped, by some foreign material which has been forced between them. If the gears are sprung the noise will be irregular.

As a perfectly quiet axle can be obtained only by correct adjustment, the first thing to be sure of is that the differential axis and the pinion axis are in the same plane. Any variation of this will throw the strain or load on either the heel or toe of the tooth. For instance, if the pinion axis is above the gear axis the strain will be thrown on the heel of the tooth. The meaning of heel and toe of the tooth is shown in Fig. 41 at a. The heel is the large end and the toe is the small end.

Bearings play a very important part in the rear axle and should be very carefully examined before being replaced after overhauling. A little angular play in single row ball-bearings is not harmful, but they should have no radial movement. Angular movement is the same movement as in a ball joint, and radial movement is from the center outwards towards the outer ball race; and if this second movement is very pronounced, the bearings should be renewed. Sometimes the balls will chip or score, as also will the races. This is another reason for renewal. Be sure the bearings spin freely and

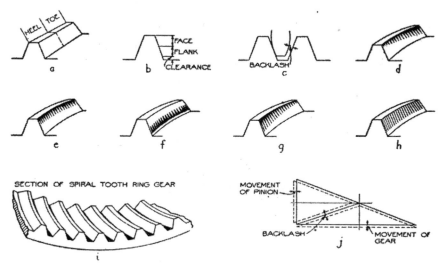

Fig. 41.   Tooth Markings for Rear Axle Gear Adjustment

are free from grit. A good plan is to wash them in gasoline, then dry them by air pressure, grease them, and then wrap them in paper or cloth until they are needed. Cleanliness is imperative in any overhauling job. Some axles have roller bearings. Examine the rollers for chipping or pitting, and the cups for hollows, and if either of these conditions is found, replace with new parts. Lastly, before replacing any bearings, cover them thoroughly with grease or vaseline.

Ring Gear Installation. If a new ring gear is to be installed, be sure that there are no burrs or chips anywhere that will prevent the ring gear from seating. Before riveting, bolt the ring gear on to the differential housing in at least three places, with bolts that are a good fit in the holes and this will prevent the riding or shifting of the ring gear, and test for alignment. Place the differential housing, with the halves bolted together and using the differential hubs as centers, in the lathe and revolve it at a good speed, and watch the ring gear. It should not run eccentrically at all and should not weave more than .005. If there is a pronounced running out of true, the ring gear must be taken off the housing, and the face of the differential housing, against which the ring gear seats, should be machined true. When this has been corrected, the ring gear may again be bolted on housing and tried as before, and, if found correct, the riveting may be commenced. Do not remove the bolts until all the other holes have been filled with rivets and the ends riveted over very tightly.

Bevel Pinion Gear Installation. When installing a new bevel gear, be sure that the key does not stand up too high in the shaft,

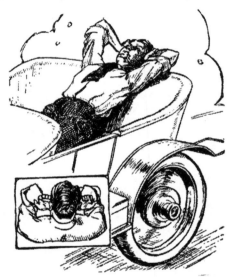

Fig. 42. Listening for Rear-Axle Noises

as this will prevent the gear from being driven on the shaft to its correct position, and it will also make the gear run out of true. An allowance of .002 inch is permissible in this alignment.

A hum will often develop in an axle that was previously quiet. If the hum sounds plainer while the engine is pulling, the gears are too deep in mesh; and if the hum sounds plainer when the car is rolling but the engine not pulling—"on the coast," as it is termed —the gears are not meshed deep enough. Of course, this is governed more or less by the condition of the gears.

A good way to listen to rear axle hums out on the road is to lay back over the rear end of the car, Fig. 42, with the head against the top of the seat and projecting over slightly, and with the hands cupped in front of the ears, so as to catch every noise that arises. The larger sketch shows the general scheme, the small inset giving the method of holding the hands. When the sound arising from the axle is a steady hum, the gears are in good condition and well adjusted. If this sound is interrupted occasionally by a sharper, harsher note, it may be assumed that there is a point in one of the gears or on one of the shafts where things are not as they should be. By trying the car at starting, slowing down, running at various speeds, and coasting, this noise can be tied to something more

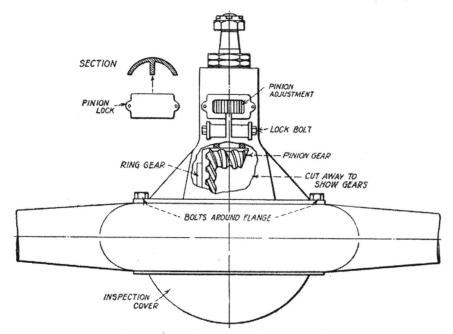

Fig. 43.   Fore-and-Aft Adjustment of Bevel Pinion

definite, some fixed method of happening. In advance of actual repair work, including tearing down the whole axle, the gears can be adjusted. This can generally be done from outside the axle casing and without a great deal of work. If the adjustment makes matters worse, it can be reversed, or if it improves the situation, the adjusting can be continued, a little at a time, until the noise gradually disappears.

Points of Adjustment. There are three points of adjustment for the meshing of the ring and bevel drive pinion gear in the rear axle. One adjustment is for the movement of the bevel pinion in a fore- and- aft direction and is located at the front of the rear axle housing and just behind the rear universal joint. Sometimes this adjustment is covered up inside the housing as shown in Fig. 43 and is held in place by a locking device that bolts on the outside of the housing. In other types, the adjusting nut is screwed directly into the end of the housing. The other two adjustments are located at the side of the differential unit and allow the movement of the ring gear in a left-hand or a right-hand direction. This adjustment moves the ring gear into deeper mesh with the bevel pinion.

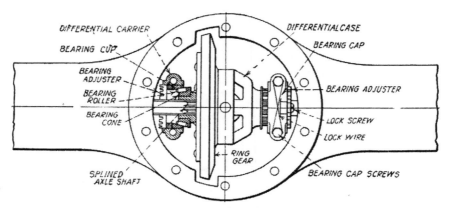

Fig. 44.   Ring Gear Adjustment

The position of this adjustment is shown in Fig. 44. In some axles the differential ring gear and bevel pinion can be removed as one unit and this makes it a very easy matter to make an adjustment after the axle has been overhauled. This unit is often called the differential bridge.

Rear Axle Gear Adjustment. The method used in adjustment by most repair men is to set the teeth of the gears so that they are flush at either the large end or the small end, with a clearance between the teeth of about .005 or .010. Although this is very satisfactory, it does not insure a perfect adjustment and should be forgotten if a perfect adjustment is wanted. A better method, after the axle has been placed under the car, is to paint the ring gear with white paint after the gears have been placed as near as possible to the position stated above. Be sure that the brakes are equalized, so that an even load may be obtained on the rear wheels. Start the engine and put the high gear in mesh. Now put on the brakes, and this will wipe the paint off the teeth. If the marking shows the same as pictured in Fig. 41-h, it is quite correct; and the axle will be found to be perfectly quiet, providing the gears are not worn. In the illustration the teeth are shown in full contact. This is the ideal condition and will give perfectly quiet operation.

If the contact is as shown in Fig. 41-d, that is, heavy on the toe of the teeth, the ring gear must be moved away from the pinion. If gears were left to run in this way, the toe would eventually break off.

Should the markings show as in Fig. 41-e, the bevel pinion or gear should be moved in deeper mesh. It will be noticed that the bearing or contact is the length of the tooth, but it does not go down deep enough, and the same thing applies to Fig. 41-f, only in this case the mesh is too deep. Remedy by moving the pinion away or out of mesh.

Fig. 41-g shows the marking for too much backlash. To correct this, move the ring gear towards the bevel pinion, but be sure you have a slight amount of play between the gears before running the car. Gears set up in this way, if run, will break off at the heel of the tooth.

If after continued adjustment, the contact still shows as in Fig. 41-g, the axle must be sprung or the gears are machined incorrectly. The axle should never be left to run with the heavy contact on the heel of the gear.

Fig. 41-j shows two cones, which represent the gears, and illustrates the difference obtained by backlash. In adjusting rear axles it should always be done if possible by moving the ring gear, but of course it is sometimes necessary to move both ring gear and bevel gear a little according to the marking obtained.

A car may be tested a short distance without grease in the housing in case of further adjustment, but do not forget to fill the axle housing before delivery. If, when driving, the axle develops a thump, it is a case of either a broken or chipped tooth on either gear, or the ring gear is sprung out of true. The foregoing method can and should be used in the adjustment of any rear axle of the adjustable type.

Rear Axle Shafts Breaking. There are three reasons for the continual breaking of axle shafts: (1) a fierce or grabbing clutch;

(2) worn bearings at the wheel end of the axle shafts; and (3) axle housing out of line with axle shaft bent. The second trouble is not so prevalent in axles that have a ball or roller bearing which revolve in their own cups or races, but it is of frequent occurrence where the bearing comes into direct contact with the axle shaft. The bearing wears into the axle shaft and the shaft—being out of true alignment —causes a twisting motion. The shaft becomes gradually crystallized and breaks off. The cure is to keep the bearings well lubricated and at the first signs of wear the axle should be replaced.

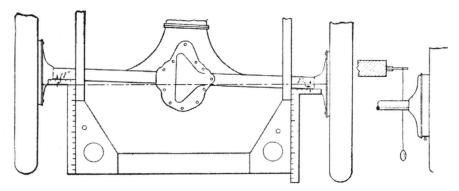

Fig. 45.   Method of Checking up Rear-Axle Alignment with Square and Plumb Bob

Wheel bearings of any kind should always be carefully adjusted and well taken care of. Frequent inspection and cleaning is half the battle towards long life and service of the bearing. To make an inspection for bearing play, jack up the rear wheel and push and pull on the wheel. This motion will show if there is any side play on the bearings. Take hold of the top and the bottom of the wheel and use a rocking motion, which will show if there is any up and down play. If there is, the bearing should be renewed. Should the axle housing be out of alignment or the shaft bent, it will break in service eventually.

Lining Up Axles. In such a repair, however, the main thing is to get the rear axle lined up correctly, which is not an easy job. This may be done in the following manner: Get the car standing level on a nice clean smooth floor; hold a large metal square with a plumb bob hanging down over its short edge against the side of the frame. Move the square forward until the line just touches the rear axle at some set distance out from the frame, say 3 inches, as shown in Fig. 45. Then notice the distance this line is forward from the rear end of the frame. In the sketch it is 16 inches. Transfer the square and plumb bob to the other side and repeat. Here it will be found that the distance from the rear end of the frame is either more or less. In the sketch it is shown at 18 inches; so the difference, 2 inches, shows that the axle is out of alignment that much or half that, 1 inch at each end.

This axle is straightened by loosening the spring clips and pushing one side back the distance apparently needed, then fasten-

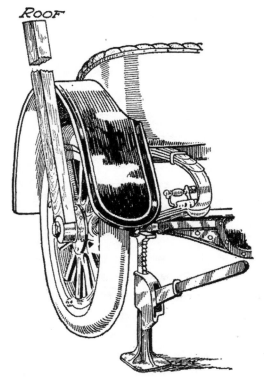

Fig. 46. One Way of Straightening Rear
Axle Quickly

ing tightly and checking up. If not correct, try again, using judgment as to which side should be moved. When finally satisfied that the rear axle is square with the frame, it is well to check this against the center-to-center distance of the wheels on each side. This is done by setting the front wheels exactly straight and then measuring from the center of the right front to the center of the right rear wheel. Then go over to the other side and measure the center-to-center distance of the left wheels. The two axles should agree exactly. If they do not, the rear axle presumably needs more adjustment for squareness.

Taking Out Bend in Axle. A simple method of repairing an axle which has been bent, but a method which is only temporary in that it is not accurate enough to give a job which could be called final, is that indicated in Fig. 46. The axle was bent when the hub struck an obstruction in the road, and it had to be straightened immediately. A short length of 2×4 timber was cut to be a tight fit between the upper side of the hub cap and the roof beam. Then

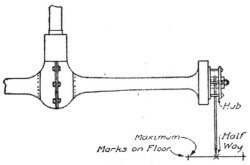

Fig. 47. Diagram Showing Method of Checking Up Ford Axles

a jack under the axle at the point of the bend was raised. As the jack raised the axle, and the wood beam held the hub down, enough pressure was exerted to force the axle to give at the bend and return as nearly as possible to its original straightness. It was a quick and easy repair of the rough and ready order, which served when time was worth more than anything else; but it is a method which would not be advised or recommended when there was sufficient time to properly straighten the axle.

Bent Ford Axles. Many cases of Ford bent rear axles can be fixed without taking down the whole construction. The principal point is to find out how much and which way the axle is bent. By removing the wheel on the bent side and placing the rig shown in Fig. 47 on the axle end, the extent of the trouble can be indicated by the axle itself. The iron rod is long and stiff, with its outer end pointed, and is fastened permanently to an old Ford hub. The rig is placed on the axle and held by the axle nut, but without the key, as the axle must be free to turn inside the hub. With the pointed end of the rod resting on the floor and with high gear engaged, have some one turn the engine over slowly, so as to turn the axle shaft around. As it revolves, the hub will be moved, and the pointed end on the floor will indicate the extent of the bend. By marking the two extreme points and dividing the distance between them, the center is found. Then a rod can be used as a bar to bend the axle, until the pointed rod end is exactly on the center mark. A

Fig. 48. How Spring Clips Are Replaced
in Emergency
Courtesy of "Motor World"

little practice with this rig will enable a workman to straighten out a Ford rear axle in about the time it takes to tell it.

Cracking Noise in Rear Wheel as Clutch Takes Hold. This trouble may be caused by the wheel being loose on the axle and the key may not be a good fit between the keyway in the axle or between the keyway in the wheel hub. It should be corrected as soon as possible or the key may be sheared off or the keyways in both axle and hub will be badly worn necessitating a new axle and wheel hub. If the rim is not tightly bolted to the rim, it will move as the wheel starts to turn and will also cause a cracking noise.

Rubbing Noise in the Rear Wheel. This noise is usually more pronounced as the car rounds a corner or a curve and is caused by the wheel being loose on the axle or by end play in the axle shaft. The noise is made by the brake drum rubbing against the shield at the back of the wheel. To cure this the wheel should be tightened on the axle or the rear wheel bearings examined and adjusted according to the trouble found.

Repair for Broken Spring Clips. The springs are held down on the axles by means of spring clips, which are simply U-shaped bolts with the inside width of the U equal to the width of the spring. Occasionally, these will break when they cannot be replaced or new ones forged. Under such circumstances, a repair such as used by one man, shown in Fig. 48, will always get the car home or to a garage where a better one can be made. This method of repair consists of a pair of flat plates, one above the spring, the other below the axle, with holes drilled in the corners to take four long carriage bolts, which happened to be handy. The plates were put on, bolts put in and tightened up, and the car was ready to run. Although an I-section axle is shown in the illustration, this method of repair would work just as well on a round axle or on an axle of any other shape. Before making this repair, be sure that the spring center bolt is not broken and after making the repair, be sure that the nuts are tight on the bolts so that the axle cannot move. Also, be sure that the axle is in the correct position and that the wheels are in line before starting to drive the car again because, if these conditions are not correct, it will cause excessive wear on the tires.

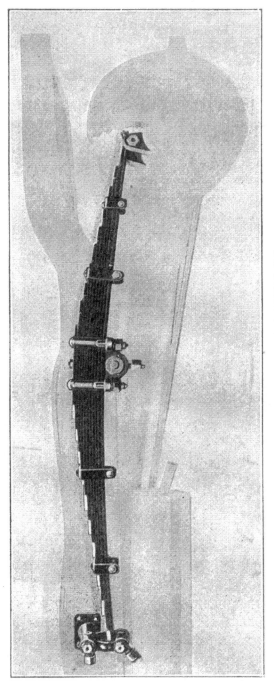

CANTILEVER SPRING ON BUICK CAR
Courtesy of Buick Motor Car Company

BRAKES

The brakes of the automobile are the most important part in the safe operation of the machine but they are the most neglected part of the automobile. The reason for this neglect may be laid to two things: namely, the fact that the adjustments are hard to reach and that a good many drivers do not know how to adjust them. The brakes should be kept adjusted and equalized at all times and frequent inspection should be made in regard to the conditions of the brake lining. Just as soon as the linings show wear and there is no adjustment left on the rods and other parts of the brake mechanisms, they should be renewed at once.

Function of Brake. Next to power, applied through the correct form of gearing, and its final suitable drive to the road wheels, nothing is of more importance than the ability to stop the vehicle at will. One medium through which this is done, and which ordinarily suffices, is the shutting off of the source of power—in this case, the closing of the throttle which feeds the gas to the cylinders. This will not always suffice, however, for the ordinary car possesses the ability to run at a speed of 40 miles per hour or upward, and weighs from 2000 pounds (one ton) upward to 4000 pounds (two tons). This combination makes for a large force of inertia, which will result in the car running for many yards, even hundreds of yards, after the power is shut off. It is for this reason that we must have a mechanical means of absorbing this inertia, or of snubbing the forward movement of the car. This is the function of the brakes, as fitted to the modern car.

The term "use the engine as a brake" means to allow the compression in the cylinders to act as a brake on the momentum of the car. The engine may be used as a brake when descending long hills in the following manner. When the top of a steep incline is reached, the car is stopped and the change speed lever placed in second or even first speed position and the car started slowly and allowed to roll down the hill against the compression of the engine. The brakes should be used only to snub the speed of the car should it roll too fast. This not only prevents the overheating of the brakes and brake drums but also reduces the wear on the brake linings. When using the engine as a brake, the switch should be placed in the off position.

Classification. Brakes are usually divided into two classes— the internal expanding and the external contracting types. The internal expanding consists of a pair of brake shoes, sometimes of metal. Often a metal shoe is covered with some kind of fabric lining, the whole expanding against the inside of the brake drum when the brake is operated. The external contracting type consists of a steel band to which is riveted a lining of some non-burning material and when the brake is operated, the band is contracted around the drum. Both of the foregoing types are divided into two classes according to the type of action used in the operation of the brakes and are as follows—cam type action, toggle type action, and scissors type action.

Brakes are also divided according to their location such as the propeller shaft or transmission brake and the axle brake. In the shaft type an external contracting band is used which is attached to the transmission with a drum which, in turn, is attached to the propeller shaft. When the brake is operated, the band contracts around the drum and checks the speed through the propeller shaft and rear axle parts. The transmission brake has quite a number of advantages: (1) reduces the weight; (2) gives more time between the renewal of the lining, because the brake is away from the road grit and mud; and also, because of the wider braking surface of the drum and band; and (3) quick and easy adjustment, the adjustment being made, as a rule, by just one thumb nut. There is no equalizing to be done.

In the axle type, the brakes are not actually attached to the axle but to the wheels through the medium of a drum.

In the last year or so a great deal of work has been done in the development of the four-wheel brakes. In this type of installation, a drum is attached to all four wheels and both the external contracting and the internal expanding brakes are used and all four brakes operated by one foot pedal. There are two systems of operation used in the four-wheel brake installation—the hydraulic and the mechanical. There is one distinct advantage that the hydraulic brake has—it is self-equalizing. The mechanically operated brake needs constant attention to keep all brakes properly adjusted and equalized. The hydraulic brakes are operated by oil and in one make by glycerine and alcohol mixed in certain proportion.

External Contracting Brakes. This class of brakes is divided into single and double acting brakes. In the first the end of a simple band is anchored at some external point, while the other, or free end, is pulled. This results in the anchorage sustaining as much pull as is given to the operating end, that is, all pull is transmitted directly to the anchorage. This disadvantage has resulted in this form becoming obsolete.

Any brake of the true double-acting type will work equally well acting forward or backward. The differential brake, Fig. 1,

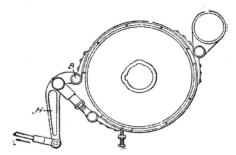

Fig. 1.   Brake on Main Shaft of
Benz (German) Car

shows this clearly. The external band is hung from the main frame by means of a stout link which is free to turn. The band itself is of very thin sheet steel, lined with some form of non-burnable belting. The ends carry drop forgings, to which the operating levers are attached. These are so shaped that the pull is evenly divided between the two sides of the band. This will be made apparent by considering that a pull on the lever H will result in two motions, neither one complete, since each depends upon the other. First, there will be a motion of the upper band end B about the extremity of the lower one as a pivot, followed by a movement of the lower end, pivot and all, about B as a second pivot point. These two motions result in a double clamping action which is supposed to distribute evenly over the surface. In order to insure even distribution, the lining is grooved, or divided, into sections.

Usually, chain-driven cars have a different brake location from a car with shaft drive. Some of these cars have three sets of brakes: one on the main shaft, one pair on the countershaft, and another pair on the rear wheels.

Internal-Expanding Brakes. While the contracting-band brake is well thought of, the internal-expanding form is rapidly displacing

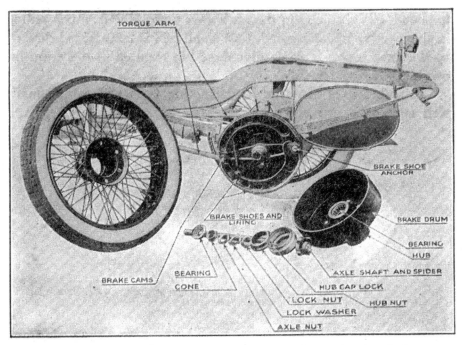

Fig. 2.   Rear Axle of Owen Magnetic Car with Wheel Removed to Show Brakes
Courtesy of Baker-R & L Company, Cleveland, Ohio

it. In Fig. 2 will be seen a modern form of the internal brake, namely, the use of both brakes as internal, but placed side by side in the same drum. This is a tendency which seems to be gaining in favor. The car is the Owen Magnetic, one of the most expensive and luxurious; so the use of side-by-side internal brakes here must be attributed to superiority rather than to a desire to save in money or in parts.

A considerable number of foreign cars, which are used in mountainous countries, show a method of cooling the brake drums by means of external cooling flanges. In some makes, even a water drip is provided for extremely hilly country.

More modern practice shows no tendency to place all of the eggs in one basket, both forms of brake being employed together and upon the same car, usually also upon the same brake drum, one set working

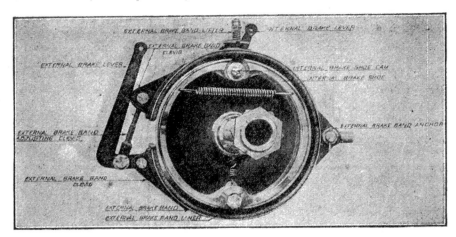

Fig. 3. Peerless Rear-Axle Brake

upon the exterior, while the other works upon the inside. In Fig. 3, which shows the rear-axle brakes of the larger cars made by the Peerless Motor Car Company, this mechanism is plainly illustrated, both the brakes being shown, although the drum upon which they work has been removed. The parts are all named so as to be self-explanatory. In this construction, the inner, or expanding, band is operated by a cam. In the brake sets put out by the Timken Roller Bearing Company, of Detroit, Michigan, in connection with their bearings, and axles, the toggle action is used, Fig. 4. The constructional drawings, Figs. 5 and 6, showing the brakes used on the Reo car, manufactured by the Reo Motor Car Company, of Lansing, Michigan, indicate that this firm is partial to the cam for brake operation, since these are used for both internal and

Fig. 4. Timken Double Rear-Axle Brake

external brakes, the internal form having a split link connected to the toggle, while the external has a link movement in contracting much like that shown in Fig. 1, which is there explained in detail.

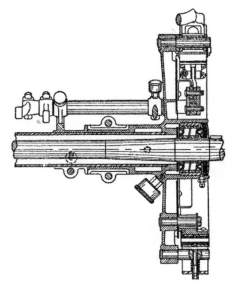

Fig. 5.   Section Showing Construction
of Reo Brakes
Courtesy of Reo Motor Car Company, Lansing, Michigan

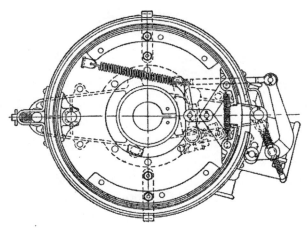

Fig. 6.   Drawing Showing Method of
Operating Reo Brakes
Courtesy of Reo Motor Car Company, Lansing, Michigan

In general, however, when both brakes are placed on the rear wheels, one external and of the contracting-band type, and the other internal and of the expanding-shoe form, modern practice calls for a cam to operate the latter, operating directly upon the ends of the two halves of the shoe, while levers operate the band so as to get a double contracting motion.

Some modern brakes may be seen in Figs. 7, 8, and 9. The first shows a system such as just described; the second shows a stiff metal shoe in both types; and the last a pair of shoes set side by side. In addition, the last-named includes a new thought in that the brake shoes are floated on their supporting pins, as shown.

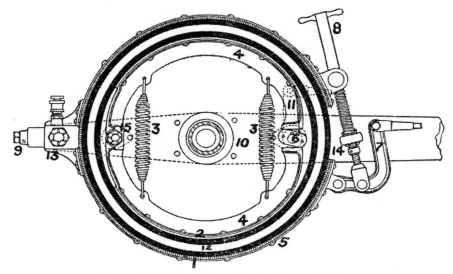

Fig. 7. Double Brake Drum Used on Locomobile Cars

This makes the bearing of the shoes certain when expanded against every portion of the drum, as the shoes can "float" until they fit exactly.

Double Brake Drum for Safety. A very important feature is pointed out in Fig. 1, namely, that of safety. Where both brakes work on a common drum, one inside and the other outside, the continuous use of the service brake (whether internal or external) heats up the drum to such an extent that when an emergency arises calling for the application of the other brake it will not grip on the hot drum, being thoroughly heated itself. The double drum allows air circulation and constant cooling.

Methods of Brake Operation. While it is generally thought that round iron rods are the universal means of brake operation, such is

Fig. 8.   Brakes and Rear Construction of Pierce Cars
Courtesy of Pierce-Arrow Motor Car Company, Buffalo, New York

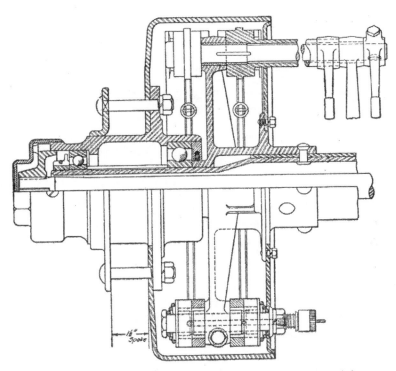

Fig. 9.   Side-by-Side Arrangement of Brakes on American Rear Axle
Courtesy of American Ball Bearing Company, Cleveland, Ohio

not the case. Many brakes on excellent cars are worked, as the illustrations show, by means of cables. This idea is even carried so far that brakes have been fitted to operate through the medium of ropes. Chains of small diameter have also been used, as well as combinations of rods, chains, cables, and ropes.

A lever-operated braking system of a well-known medium-priced car is shown in the outline sketch, Fig. 10. In this system

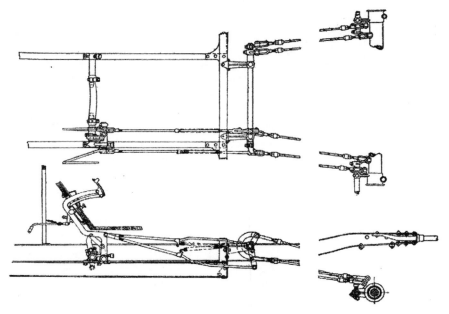

Fig. 10.   Layout of Brake-Operating System Using Cables

the forward part of each half is worked by rods moved by means of pedals, but the rear part of each half is actuated by means of cables. Cables have one advantage over rods in a situation like this—the diagonal pull with a stiff rod might, in time, act to pull the brake side-wise off their respective brake drums, the cable, being more flexible gives less danger of this.

This method of operation seems to be gaining favor because of its simplicity, which eliminates parts that add weight and gives immediate results when the parts are properly adjusted. The recent New York show revealed a surprising number of small and medium size cars with cable-operated brakes. An inspection of these cars showed a mechanical cleanliness which was lacking in many others of the same class on which an attempt was made to reduce braking rods and levers to a minimum, with consequent bent levers, bent or crooked rods, brakes worked from an angle, and other unmechanical ideas.

Fully as important as the operating means is the matter of equalizing the pull so that the same force is exerted upon both wheels at once. This action is influential in causing side-slip or skidding, which may result fatally. To equalize the force was one reason for

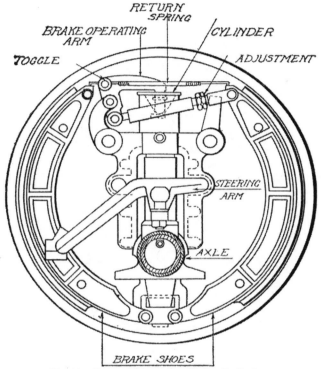

Fig. 11. Internal Expanding Hydraulic Brake

the use of cables, although the more up-to-date way is to attach the operating lever to the center of a long bar, to the extremities of which the brakes themselves are fastened. A pull on the bar is then divided into two different pulls on the brakes, the division being made automatically and according to their respective needs. This is an important point, and one that should be looked after in the purchase of a new car.

Four Wheel Brakes. Four wheel brakes are not new since they were used in Europe for a number of years, for instance on the Renault, but it is not until recent date that they have come into use in America. There are two types of brakes used in the four-wheel installations, hydraulic and mechanically operated, and both systems are used with both the external and internal type of brake.

Internal Expanding Hydraulic Brakes. The Duesenberg "Straight-Eight" uses the hydraulic brake. The advantage of this system is that it eliminates the use of rods and clevis pins, which wear and cause a great deal of rattle. It also gives a neater appear-

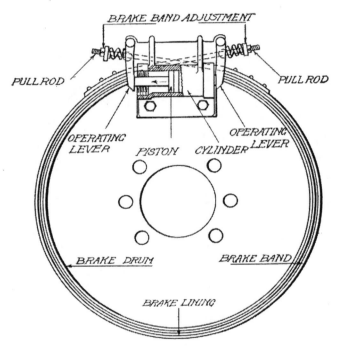

Fig. 12. External Contracting Hydraulic Brake

ance. The system also equalizes the pressure applied to the brakes and eliminates skidding. The operation of the brakes is very simple. Attached to the brake pedal is a piston which moves in a cylinder. When the pedal is depressed, the liquid is forced out of the cylinder by the piston and is lead by tubes to the brakes of the wheel. A small cylinder is attached at each of the four brakes and as the oil enters these cylinders the piston is forced to move and operate the toggle arms, which in turn operate the two brake shoes in each brake drum, Fig. 11. In the system spoken of there is a special cylinder system which takes care of keeping the system full of oil at all times and if there should be a leak in the system it is immediately noticed because the brake pedal travels farther than it should and the action causes the special cylinder to come into operation which pumps additional oil into the system to take care of the leakage. With this system of breaking, the Duesenberg Company uses a tubular axle which gives the additional strength to withstand the breaking torque or twisting strain.

External Contracting Hydraulic Brakes. The external contracting type brake is operated on the same principles as the expanding

Fig. 13.   Expanding Internal Brake Used on the Packard Car
Courtesy of Packard Motor Car Company

type. In one installation as made by the Wheel Hydraulic Brake Corporation, there are two pistons in the operating cylinders which move in opposite directions against the levers which are designed to come into contact with rods that are attached to the brake bands, Fig. 12. The liquid medium in this system is a mixture of alcohol and glycerine and is used because of its low freezing point. An Elliott axle of the reverse type is recommended for use with this type of installation.

Internal Expanding Mechanical Brakes. At the present time the mechanically operated brake seems to be the most popular for the four-wheel brake system. The Packard Company use an expanding internal brake on the front wheel. Fig. 13 shows the

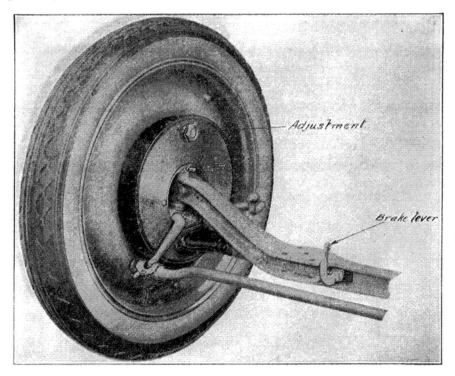

Fig. 14. Front Brake and Axle Bracket Used on the Rickenbacker Car
Courtesy of Rickenbacker Motor Car Company

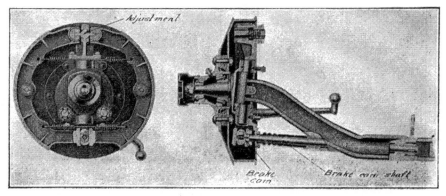

Fig. 15. Cross-Section of Internal Expanding Mechanical Brake as Used
on Rickenbacker Car
Courtesy of Rickenbacker Motor Car Company

method of operation and installation. An inverted Elliott axle is used with this equipment and ribbed brake drums are used for

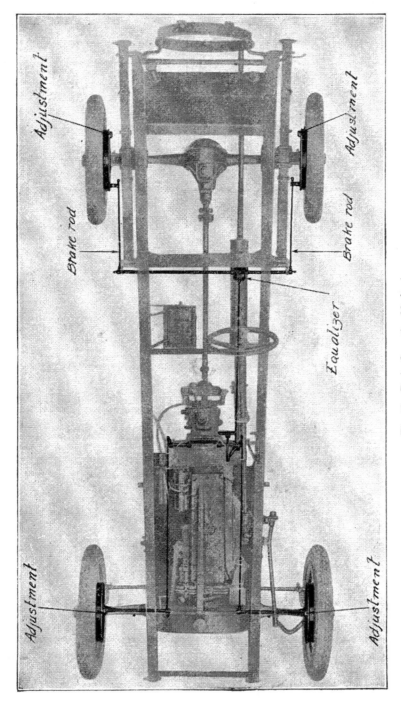

Fig. 16. Brake Operating Mechanism
Courtesy of Rickenbacker Motor Car Company

113

cooling purposes. Another excellent example of an expanding type of brake is that of the Rickenbacker car. An interesting comparison can be made between the Packard and the Rickenbacker installations. In the Packard equipment, the operating mechanism is above the front axle with a universal joint between the frame and the front wheel to allow for turning, while in the Rickenbacker, the shaft that operates the brake passes through a bracket bearing in the front

Fig. 17. Equalizing Gears
Courtesy of Rickenbacker Motor Car Company

axle, Fig. 14 and Fig. 15. The arrangement of the pull rods on the Rickenbacker is shown in Fig. 16. The adjustment and equalizing of the average four-wheel brakes is rather a complicated job but in the Rickenbacker the equalizing is taken care of automatically by sets of small gears placed one on the cross-shaft and the other on the bottom of the foot brake pedal. The gears on the bottom of the foot pedal are shown in Fig. 17. To adjust the brakes on the Ricken-backer, it is necessary to jack up all wheels from the ground. Dis-connect all of the brake rod levers on the end of the equalizer shafts and see that both equalizers work freely and smoothly. Tighten the brake adjusting nut until the wheel can be turned when pulling on the tire. There must be the same tension on all wheels. Pull up

Fig. 18. External Contracting Mechanical Brake Used on Buick Cars
Courtesy of Buick Motor Car Company

the levers until the pins can be inserted and if they do not slide in freely, the rods must be made longer until they will do so. In practically all installations the same method of jacking up and freeing the rods is used in the adjustment but, of course, each system has its own type of adjustment.

External Contracting Mechanical Brakes. An example of this type of installation is shown in Fig. 18, which shows the Buick front-wheel brake. It will be noticed that the method of operating the brake is very similar to the Packard method, which also uses the universal joint. An inverted Elliott axle is also used in this installation. The difficulty with the four-wheel brake system is in getting the correct braking when turning corners. If the front wheels were locked, the car would start to skid and the steering control would be lost. In the Buick system the outside wheel runs freely while the inside wheel is held firmly. This allows the brake to be operated when turning a corner and still be efficient. All cars with the four-wheel brake installations use some method to accomplish a like

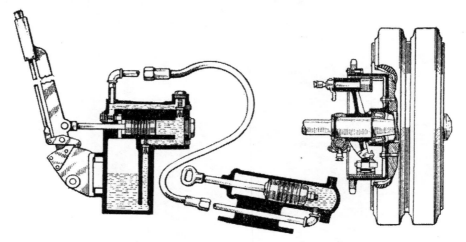

Fig. 19. Layout of Hydraulic Brakes Used on Knox Tractor
Courtesy of Knox Motors Company, Springfield, Massachusetts

result. In the Buick, the pin of the universal joint is at a different angle to the steering knuckle pin, while in the Rickenbacker, the face of the cam, which operates the brake, is made in such a way that the effectiveness of the front-wheel brake is less as the wheel is turned. The equalizers on the Buick car are of the bar type.

Hydraulic Brakes on Tractors. On the newer Knox tractors, a brake of very large size is made even more powerful by hydraulic operation. This brake is shown in Fig. 19. At the left will be seen the usual brake lever attached to a small piston in a chamber full of liquid. This chamber communicates through the medium of a valve normally held closed by a spring, with a passage above, and that, in turn, communicates with the pipes leading to the brake-operating cylinder. This cylinder has a stout rod attached to a good size plunger, back of which the liquid (oil) is introduced. When liquid is forced in, the plunger moves forward, forcing the rod out and, through connecting rods and levers, applying the brakes. These brakes, which are of the internal-expanding type, are exceptional in size and work against steel drums attached directly to the wheel spokes.

When the lever is drawn back in the usual manner, liquid is forced upward through the top passage to and through the pipes into the other cylinder, forcing the plunger to move and, through the movement of the plunger, the brakes are applied. The return of the fluid is not shown, but it is assumed that this is through a simple pipe connected from the plunger cylinder to the hand-operated piston with a check valve. Should the initial movement of the lever fail to apply the brakes sufficiently, the driver can let the lever come forward and then pull it back again, in so doing he will take into his lever cylinder more liquid from below without releasing the brakes. Then, when this extra quantity is forced through, the plunger is moved even farther forward, and the brakes applied more forcibly. The brakes are 20 inches in diameter by $6\frac{1}{2}$ inches wide.

BRAKE TROUBLES AND REPAIRS

Dragging Brakes. Probably the first trouble in the way of brakes is that of dragging, that is, braking surface constantly in contact with the brake drum. This should not be the case, as springs are usually provided to hold the brake bands off the drums. Look for these springs and see if they are in good condition. One or both of the brake bands may be bent so that the band touches the drum at a single point.

Another kind of dragging is that in which the brakes are adjusted too tightly—so tightly, in fact, that they are working all the time. In operating the car, there will be a noticeable lack of power and speed, while the rear axle will heat constantly. This can be detected by raising either rear wheel or both by means of a jack, a quick lifting arrangement, or a crane, and then spinning the wheels. If the brakes are dragging, they will not turn freely.

All that is needed to remedy this trouble is a better adjustment. For the new man, however, it is a nice little trick to adjust a pair of brakes so that they will take hold the instant the foot touches the pedal, that they will apply exactly the same pressure on the two wheels, and yet will not run so loose as to rattle nor so tight as to drag.

Brakes Do Not Hold. There are several causes for this condition. The brakes may need adjusting or relining. The brake band may be greasy owing to the oil leaking out of the rear-axle housing and getting on the brake bands. If this condition exists, remove the wheel and insert a new oil retaining washer, which is usually made of felt, in the bearing retaining nut. An excellent help in this trouble is to wind a strip of felt around the axle shaft before putting the shaft back into position. The felt should be wound so that it

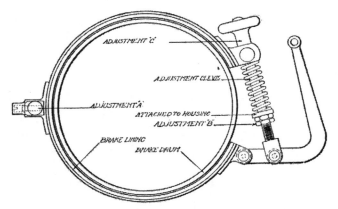

Fig. 20. Three-Point Brake Adjustment

tends to push the oil or grease away from the wheel. On the right-hand axle the felt should be bound to the left and on the left-hand axle to the right, as you face the front of the car. The brakes should be thoroughly washed in gasoline to get rid of the grease and if very badly soaked, they should be burnt off with a blow torch.

Brake Adjustment. When starting to make any kind of brake adjustment the wheels should be jacked up clear of the floor. All brake pull rods should be disconnected so that there is no tension on the brake operating levers on the wheels which might cause the wheels to drag and give a wrong impression of the adjustment. When possible, the adjustment should be made at the adjustment points on the brake band. There are as a rule three adjustments on the brake band, Fig. 20, and this type is called the three-point suspension. The first point of adjustment is at the point *A*. The adjustment screw at this point should be turned in such a way that the band is moved nearer to the wheel, allowing sufficient play for the wheel to turn freely. The next point of adjustment is at the brake band clevis or at point *B*. This adjustment brings the lower part of the brake band up closer to the brake drum. The nut should be screwed down until the band is as close as possible to the drum without the wheel dragging. The next point of adjustment is at the top of the brake clevis, or at point *C*. This adjustment brings the top part of the brake band down toward the brake drum and is the final ad-

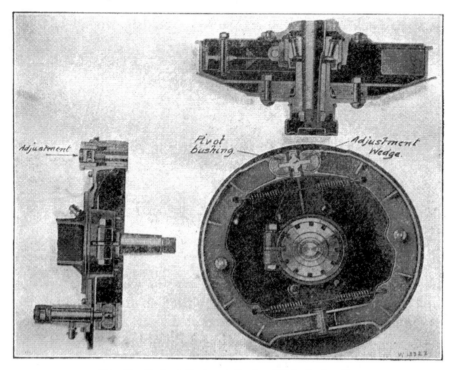

Fig. 21. Brake Adjustment Used on Rickenbacker Car
Courtesy of Rickenbacker Motor Car Company

justment as far as the band is concerned. The wheel should turn perfectly free after all adjustments are made. Connect up the different pull rods throughout the system. It is important that each wheel has the same amount of breaking tension or else, when the brakes are applied, the wheel will lock and cause the car to skid. To avoid this the brakes must be equalized and the following method is recommended. Have an assistant press down on the pedal or lever gradually and try each wheel at intervals as the pressure is increased. If the brakes are equalized, the wheels will be locked at the same point of pedal movement. Find the point at which one wheel starts to drag and adjust the other wheels so that they will start to drag at this point. There is a turn buckle on the brake pull rod and as the buckle is turned the rod will be made shorter or longer and the correct adjustment obtained. A few turns will shorten the rod a fair amount and the equalizing can be quickly and easily done in this manner. The Rickenbacker Company have a very ingenious method of adjustment and equalizing on their car. The brake shoes are adjusted by a wedge which increases or decreases the distance

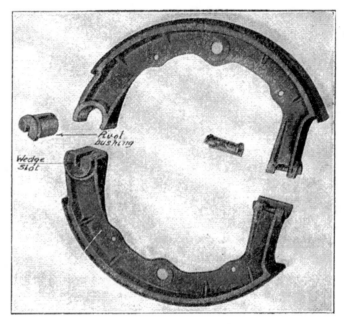

Fig. 22.   Adjustment Wedge
Courtesy of Rickenbacker Motor Car Company

between the brake shoe pivots. The wedge is drawn in by the turning of a nut which is held in position by a spring and ball, Fig. 21. As the adjustment is moved, a click is heard. By counting the number of clicks and giving each brake the same number of clicks, they are taken up equally. Fig. 22 shows the wedge shape pivot into which the wedge is drawn. The equalizing is taken care of by gears. The brake arrangement of this car makes the adjustment of four-wheel brakes a simple matter. There will be no difficulty in adjustment if the foregoing directions are followed closely.

Dummy Brake Drum Useful. Where a great deal of brake work is to be done, particularly in a shop where the greater part of the cars are of one make, and the brakes all of one size, a great deal of time and trouble can be saved by having a set of test drums. An ordinary brake drum with a section cut out so that the action inside may be observed is all that is necessary, except that it should be mounted suitably. As shown in Fig. 23, it is well to fit a pair of handles to the brake drum to assist in turning the drum when the adjustment is being made. The real saving consists of the work which is saved in putting on and taking off the heavy and bulky

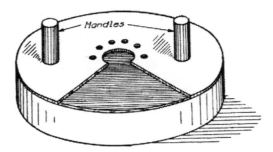

Fig. 23. Dummy Brake Drum for Adjustment Work

wheel each time when the adjustment is changed. The test drum is put on instead, and it is easily and quickly lifted on and off.

To Stop Brake Chattering. It is claimed that the chattering of brakes is caused by having the brake lining, particularly of internal hand brakes, extend over too large a portion of the circumference of the drum. The result is that with a well-adjusted system, as soon as force is applied, the lining close to the operating cam and that on the opposite side close to the pin on which the brake shoes are pivoted jumps against the drum and then away from it. This jumping of the brake shoe, which is the result of too much lining, is what causes the chattering. If the lining is cut away for about 30 degrees on either side of a line drawn from cam to pivot pin, as shown in Fig. 24, it is said that this chattering will stop immediately. If further trouble of the same kind results, bevel off the outside ends of the lining at the two 30-degree points.

A number of suggestions in the way of possible brake troubles, particularly on the side-by-side form of internal-expanding brakes, are indicated in Fig. 25. This shows a semi-floating form of rear axle with the two sets of brakes and operating shaft and levers. A number of suggestions are offered for this form, the most important of which is: "Renew worn brake lining and broken or loose rivets."

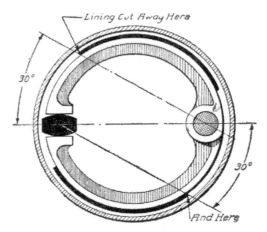

Fig. 24. Method of Eliminating Brake Chattering
on Internal-Expanding Brakes

When a brake lining is worn, the proceeding is much the same as with a clutch leather, with the exception that whereas the latter must have a curved shape, the former can be perfectly straight and

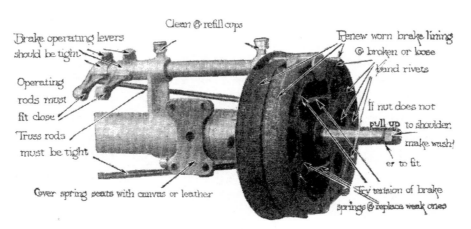

Fig. 25. Brake Troubles Illustrated

flat. This simplifies the cutting, but most brake linings are made of special heatproof asbestos composition which is made in standard widths to fit all brakes, so the cutting of leather brake bands is not often necessary.

Eliminating Noises. Many times the brake rods and levers wear just enough to rattle and make a noise when running over rough roads or cobblestone pavements, but hardly enough to warrant replacing them. The replacement depends on the accuracy with which they work, the age and value of the car, and the attitude of the owner.

Wear in the brake mechanism occurs at the clevis and clevis pins and also at the pivot pins on which the brake shoes move. The brake camshaft and camshaft bushings also wear considerably. In some cases the clevis pin hole is in a part that is expensive to renew. Fill the hole by welding, redrill the hole, and install a new clevis pin. The internal brake should operate on pins which must be renewed but the bushing in the shoe can be easily replaced. The brake camshaft can be repaired by attaching metal to the shaft by welding and turning the shaft down to the correct size. The brake camshaft bushing should be replaced if the shaft is loose after it has been repaired and tried in the bushing.

If the rod crosses a frame cross-member or is near any other metal part, and its length or looseness at the ends is such that it can be shaken into contact there, a rattle will result at that point. This can be remedied or rather deadened by wrapping one part or the other. For this purpose, string or twine can be used as on a baseball bat or tennis racket handle, winding it together closely so as to make a continuous covering. Tire or similar tape may also be utilized. When this is done, it is necessary to lap one layer partly over the next in order to keep the whole tight and neat. It has the additional advantage of giving a greater thickness and thus greater resistance to wear. If none of these remedies are available or sufficient, burlap in strips or other cloth may be used, putting this on in overlapping layers the same as the tape.

The springs should be put on in such a way as to take up the lost motion and hold the worn parts closer together. The rattle occurs when the movement of the car alternately separates and pulls together the two parts, a noise occurring at each motion. The spring should be put on so as to oppose this motion, acting really as a new bushing or pin, the pull coming first upon the spring and then upon the bushing or pin.

A squeaky brake is often caused by a rivet projecting beyond the brake lining and also by the brake lining becoming glazed.

Inspect every rivet and rerivet where necessary. In the second case, rough up the lining with a wire brush.

Brake Relining. In relining brakes it is essential that the brake lining fit the band and shoe as closely as possible. If care is not taken, it may be found that after the lining has been riveted to the band or shoe that there are flat spots in the lining. By flat spots is meant that instead of the lining following the band or shoe it is holding away from the band thus making a flat portion on the lining. If this is left, it will not allow a correct brake adjustment as the flat part will rub against the drum and cause the wheel to drag; on the other hand, if the adjustment is made to allow for this, the brake will not hold well when applied. When relining a brake

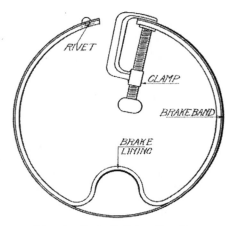

Fig. 26. Brake Relining First Step

band, the length of the lining should be found as close as possible and a little more lining allowed than is absolutely necessary so that when the lining is forced into place it will fit tight against the steel band. A method of doing this is shown in Fig. 26. Mark the position of the rivet holes in one end of the brake band lining, drill the rivet holes, and countersink the lining for the heads of the rivets. Rivet the end of the lining to the band and fasten the other end of the lining to the band with a clamp as shown in the figure. Place the surplus lining in the form of a loop as also shown. With the clamp holding the lining tight, hammer the lining into position against the brake band, Fig. 27. An inch of brake lining longer than the actual length is usually enough for the loop. Too much must not be allowed for the loop or the brake lining is apt to break when hammered into position. Drill and countersink the lining for the rivets. Do not countersink the lining too deeply so that the rivets pull through the lining when riveted over. The rivets should be placed on a piece of metal held in a vice when riveting over so that there is something solid to rivet against. The top of the riveting piece should be the same size as the head of the rivet and the band should be held squarely on the top so that the rivet will be forced firmly into the lining without any part of the head being higher than any other part. When relining a brake shoe, the lining should be a little shorter but, in this case, both ends of the

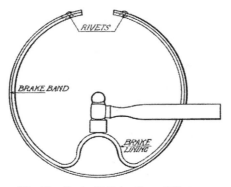

Fig. 27.   Brake Relining Second Step

lining are riveted to the shoe and the lining forced over onto the shoe. The rest of the work is the same as that given for the band job. When choosing the brake lining, be sure that it is of the correct thickness and width.

Truing Brake Drums. When both inside and outside surfaces of the brake drum are used, it is highly important that both be true. Since they do not stay that way long, the repair shop should be equipped to true them up quickly. A truing device is shown in Fig. 28, with the wheel and brake drum in place on it. One feature of the device is that brake drums need not be removed from the wheel. The device consists of a metal base having a strong and stiff wooden pier with a horizontal arm the exact size of the axle end mounted on it. The wheels are placed on the arm and rest on it the same as on the axle when on the car. The tool is double, with two ends, one of which cuts the inside surface of the drum, while the other cuts the outer surface. At the center this tool is attached to a heavy casting, bored out to slide over the shaft and with a key fitted into a keyway in the shaft to prevent the tool from rotating. The end of the arm is threaded, and a large nut with two long rams

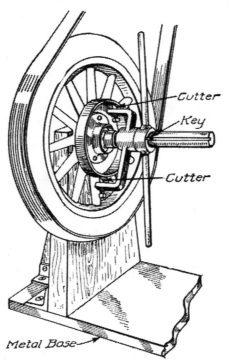

Fig. 28. Apparatus for Truing Inside and Outside of Brake Drum in Place on Wheel
Courtesy of "Motor World"

is screwed up against the tool at the start, and then it is used to feed the latter across the work.

This is subject to a number of modifications to fit it to the various sizes and shapes of brake drum. Another method is to use the lathe, provided the shop is equipped with a lathe large enough. By making a mandrel the same as the axle spindle and having a pair of dummy bearings to place on it, the brake drum can be slipped on to the mandrel, and the whole put right into the lathe. The surface, either internal or external or both, can then be trued up exactly as if the drum were on the axle.

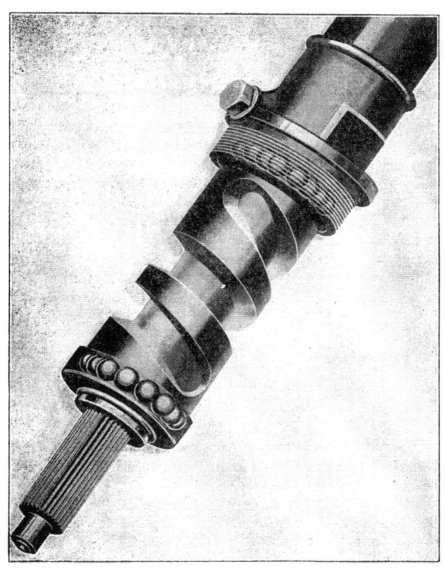

ROSS WORM STEERING GEAR

STEERING APPARATUS

STEERING GEARS

The mechanisms by which steering is effected are among the most important features of a car, if not actually the most important. The truth of this statement will be realized when attention is called to the fact that safe steering is the final requisite that has made the modern high speeds possible, for without safe and dependable steering gears, no racing driver would dare to run a machine at a high rate of speed, knowing that at any minute the unsafe steering apparatus might shift the control, thus allowing the front wheels to waver and the car to run into some obstruction by the roadside.

The same argument applies in an even greater degree to the case of the non-professional driver, who wants to be on the safe side even more, perhaps, than do the dare-devils who drive racing cars. Nearly all of our roads are curved and, to make all of these turns with safety, the steering gear must be reliable. Again, in mountainous country where there may be a sheer drop at the roadside of hundreds of feet, it becomes necessary that the steering mechanism be very accurate and that it obey, at once, the slightest move on the driver's part. To secure this accuracy, there must be no lost motion or wear of the interrelated parts.

These things mean that the whole steering mechanism must be safe and reliable; strong and accurate; well made and carefully fitted; well cared for; and finally, the design and construction must be based on a theoretically correct principle, for otherwise the mechanical refinements will have been wasted. Perhaps it will be more logical to treat the mechanical requirements first by showing how the present type has been evolved from the failures of earlier forms.

General Requirements. In turning a corner a car follows a curve, the outer wheels obviously following curves of longer radius than do the inner wheels and, therefore, traveling farther. In straight-ahead running, the wheels run parallel at all times and travel the same distance. These two facts are the basic ones which make the steering action so complicated: First, that on straight-ahead running the wheels must travel the same distance; and second, that on turning curves the outer wheels, whichever they may be, must travel a greater distance.

This double requirement leads to the usual form of steering arrangement, called after its inventor, the "Ackerman." It was Ackerman who brought out the first vehicle in which the front wheels were mounted upon pivoted-axle ends, these ends being pivoted on the extremities of the central part of a fixed axle, while the pivoted ends carried one lever each. These levers were connected together by means of a cross-rod, while at one end another rod was attached, which was used to move the wheels. By moving this latter rod, both wheels were compelled to turn about their pivot points, since the cross-rod joined them together, and if one moved the other had to move also. This was Ackerman's substitute for the fifth wheel which had been used up to that time.

Inclining Steering=Knuckle Pins. The position of the steering-knuckle pins has a great deal to do with the ease by which the steering gear is operated and also the amount of service that is obtained from the tires on the front wheels. The ideal position for the steering-knuckle pins is at the center of the wheel vertically, which would give the maximum turning movement and prevent all chance of wheel wabble but is not suitable for general use. There is a tendency toward centralizing the pin with the introduction of the four-wheel brakes. The pins are placed as near to this position as possible, however, and are usually within six inches.

The axle-spindle center may be brought close to the wheel hub by means of a double yoke, but this was tried and abandoned as too cumbersome for the results effected. A method of placing the steering pivot in the center of the wheel was also developed. In this case the pivot was enclosed in a hollow hub; but as this made the pivot, which is liable to wear, inaccessible, it also was abandoned. However, later tendencies point toward a revival of this construction.

The result is that today we are using a form which, though far from being ideal, fulfills every practical requirement. This form is usually constructed as in Fig. 1, which shows a skeleton plan view of an automobile. In this, the line AB represents in length, position, and direction, the front axle of a car, while ML represents in a similar manner the rear axle. A and B also are the pivot points for the axle-stud ends or, as they are more commonly called, the steering knuckles or steering pivots, which are represented by the lines *AD* and *BC*.

The rear (or front, as the case may be) ends of the steering knuckles are joined by the connecting rod *DC*. The Ackerman construction is such that the center lines of the steering arms, or levers, *AD* and *BC*, prolonged, must pass through the center point of the rear axle at *K*; the reason for this is that the front wheels are supposed to turn about the center of the rear axle as a center.

The angle at which the steering arms are placed to the steering knuckles depends upon the angle which the lines drawn through

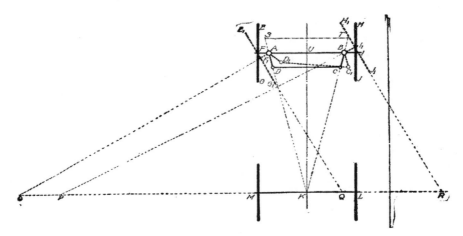

Fig. 1. Diagram of Steering Connections

AD and **BC** form with the steering knuckles. This is a rule that should be remembered when the steering arms are being corrected or straightened after trouble with the steering gear.

Camber. Camber is the angle at which the wheel is placed to the road surface and is sometimes accomplished by tilting the top of the steering-knuckle pins inward so that if a line were drawn through the center of the pin it would strike the same point on the road as would a line drawn through the center of the wheel. Camber is also obtained in some cases by the tilting downward of the axle. Fig. 2 illustrates the camber.

Caster. Caster also has a great deal to do with easy steering and its effect on the front wheels is to bring them back automatically to a straight-away position after making a turn. It is obtained by tilting the steering-knuckle pins back about three degrees and if a line was drawn through the center of the pins it would strike the road at a point a short distance ahead of the point at which the

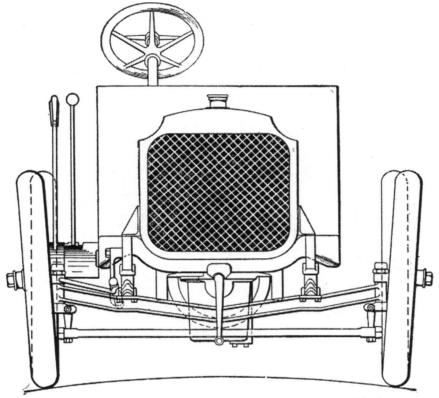

Fig. 2.   Front Elevation of Car, Showing Camber of the Front Wheels

center line drawn through the wheel would strike, Fig. 3. It will be seen from the foregoing that the steering-knuckle pins are never in a vertical position.

Action of Wheels in Turning. If the wheels are supposed to turn through an angle, the action of the above arrangement will be seen. Suppose the steering gear (not shown in Fig. 1) is turned so as to move the steering lever AD to the new position, shown dotted at AD_1. This movement will also move the other lever BC to a new position, shown dotted at BC_2. It will be noted in this position that the angle through which the right-hand lever BC has swung is not as great as that through which the left-hand lever AD has moved, although the two levers are attached together by means of the cross-connection DC.

The wheels are mounted upon the extremities of the steering knuckles at F and I; EG represents the left wheel, and HJ the right wheel. These turn about the pivot points A and B, with the movement of the steering knuckles to the new positions, shown

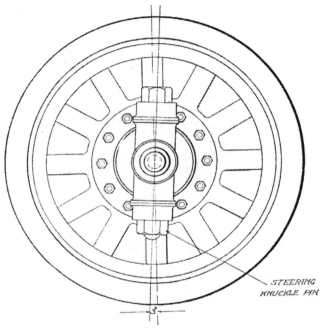

Fig. 3.   Caster

dotted at $E_1F_1G_1$ and $H_1I_1J_1$. In this position, prolongations of the lines through the pivot point and the center of the two wheels will meet the rear-axle center line prolonged at separate points as OP, the two lines converging slightly. This same convergence may be noted by prolonging the center line of the two wheels E_1G_1 to Q and H_1J_1 to R. This divergence means that the two wheels are turning on curves of different radii, and since the outer wheel H_1J_1 shows a longer distance from its center line prolonged to the rear-axle line $OPMKL$ than does the inner wheel, that is, has the longer false radius, PI_1 being longer than OF_1, it follows that the turning action will be correct.

GASOLINE AUTOMOBILES

This is somewhat complicated and rather hard to follow, but the figure seems simple and should be examined closely, even drawing it out step by step, as outlined above, for the purpose of making the steering action clear. Laying this out for one's self will bring out the reason why the steering knuckles do not move through the same distance and thus bring about a different movement of the wheels.

Steering Levers in Front of Axle. That the final movement of the wheels will not be changed if the levers, Fig. 1, are laid out

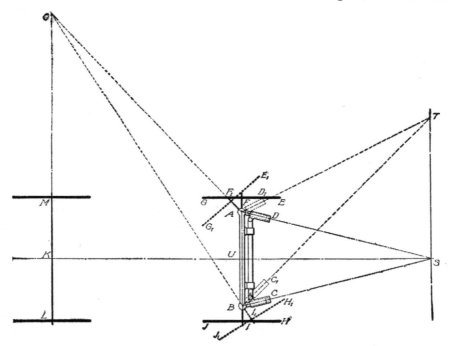

Fig. 4. Patented English Steering Device, Said to be Theoretically **Perfect**

in the same way but in front of the axle will be evident by prolonging the levers to S and T, respectively, making the lengths AS and BT the same as the former lengths AD and BC. Connecting the two by the rod ST completes the front arrangement, which is seen to give the same results as the other. The choice of a front or rear location depends upon certain things, such as the safety of the cross-rod, etc., which will be brought out later on. Some machine manufacturers even go so far as to fit both front and rear levers to the same machine.

While shifting the lever from rear to front in Fig. 1 does not change the result at all, in Fig. 4 it does. In this construction, known as the Davis, the steering levers are set in front, but taper inward instead of outward, so that their center lines prolonged meet the center line of the car prolonged at a distance from the front axle equal to the distance between the front and rear axles, or equal to the wheel base.

In addition, the connecting rod is carried in guides placed on the front of the axle, so that its path of travel is always parallel to the front axle. Consequently, the levers must be made slotted or telescopic. The result of this combination of movements is an absolutely correct angle to both wheels for any angle of lock. This can be explained by a reference to the diagram.

In Fig. 1 the prolongations of the wheel center lines, or radii of turning, do not strike the center line of the rear axle—about which they are supposed to turn—at a common point, the difference being the amount they are out of true, viz, the distance between the points O and P. If Fig. 4 be lettered to correspond with Fig. 1, the prolongations of the knuckle center lines AF_1O and I_1BP in Fig. 1 becoming the two converging lines AF_1O and I_1BO meeting at the point O on the center line LMO of the rear axle prolonged. This is as it should be and shows the case of correct steering and turning.

In this case, all four wheels are turning about the point O, the two rear wheels with the radii OM and OL, and the two front wheels with the radii OF_1 and OI_1, respectively. This gives a theoretically correct case in which all wheels will round any curve as they should and not slip or slide around, damaging the tires in the process. The Davis type of steering gear, it may be remarked, is not in general use, its construction adding a number of parts to the more usual form, shown in Fig. 1, which gives close enough results for average use.

Like the sliding-gear transmission, a steering gear is a form of mechanism which, although used on nearly all automobiles, is, from a theoretical and mechanical standpoint, far from what it should be.

General Characteristics of Steering Gears. *Standard Types.* The movement or deflection of the front road wheels is obtained by a crosswise movement of the tie rod which links the steering-knuckle levers attached to the wheels. This tie rod, sometimes referred to as the cross-connecting rod, is actuated by the drag link *GF*, Fig. 5, which is pivotally mounted on the steering-knuckle lever *L*. The drag link has a linear movement along the frame and is parallel with it.

The drag link is also pivotally mounted at the ball arm of the steering gear C, and when the drag link is moved forward or backward by movement of the ball arm, the tie rod is moved at right angles, deflecting the wheels. The drag link has a semi-rotary motion; that is, its upper end is turned through a part of a revolu-

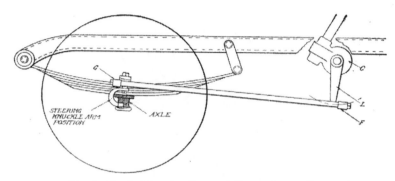

Fig. 5. Typical Steering Gear and Connections to Front Axle

tion while its lower end, to which the drag link is attached, swings through a fairly large arc, according to the capacity and design of the steering gear.

As the ball arm swings through its arc, the drag link attached to it rises and falls slightly, the movement being indicated by the dotted lines in Fig. 5. The partial circular motion in a vertical plane is converted from the rotation of the steering gear in a horizontal plane by several methods. The gear shown in Fig. 6 is known as the worm and sector type, which is illustrated in Fig. 5.

In Fig. 6 the steering column or post *CD* carries a worm *F* which is in mesh with the gear E. Rotating the column *CD* in the direction indicated by the arrows, or counter-clockwise, will result in the worm turning in the same direction. The gear *E* will rotate on its horizontal shaft in a downward movement, as shown by the arrow, and as the ball arm, or lever, is attached to the shaft, the member *L* will move backward, or to the left, as shown by the arrow intersecting the ball. With the worm type the two gears are usually in two different planes at right angles to each other, one vertical and the other horizontal. This is an advantage in that it lends itself readily to the construction of a simple steering-gear system. Thus the post is in a vertical or modified vertical line, as is also the motion of the steering arm, and the consequent movement of the steering rod is more or less confined to a vertical plane. With the worm and gear this is obtained in a simple manner. The gear-

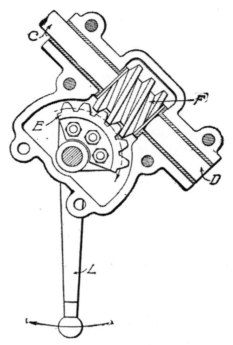

Fig. 6.   Worm and Partial Gear of Typical
Steering Gear

shaft is in a horizontal plane passing through the center line of the worm. If the worm rotates in a direction which approximates a horizontal circle around a vertical axis, the worm gear will turn in a vertical plane about a horizontal axis. A lever attached to the end of this shaft will, consequently, move in the desired plane— the vertical one mentioned before—and the desired requirements are met.

The conversion of rotary motion in a horizontal plane to partial rotation in a vertical plane is shown in Fig. 7, the action here being slightly amplified. The steering, or hand, wheel A with spokes B is turned to the left, turning the steering column C (a hollow tube) in the direction indicated by the small arrow. *D* is the steering gear with its ball arm *E*. The turning of the hand wheel moves the ball end *F* and drag link backward. The front end of the drag link is attached to the steering knuckle *M* at *H* and turns about the center line *KL* of the steering knuckle *J*, the end turning through the arc *HI*. The lever *M* is attached to the knuckle *J* and turns with it. Its end turns through the arc *OP*, moving the tie rod *OQ* to the right and turning the other knuckle in the same way and direction. *YY* are the spring pads and *ZZ* the tapered roller bearings supporting the road wheels.

There are two types of steering gears—reversible and irreversible, and three general forms of steering gears known as the worm, the

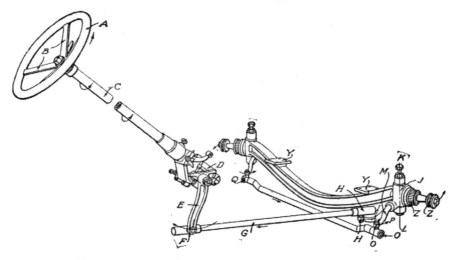

Fig. 7. Steering Mechanism and Front Axle of Pierce-Arrow Car
Courtesy of Pierce-Arrow Motor Car Company, Buffalo, New York

bevel, and the spur-gear type. These may be subdivided, which might lead one to assume that there are a dozen or more different forms. The mechanical lever has been discarded because of its tendency to impart all road shocks to the driver; it is fully reversible at all times. Irreversibility is employed because it transmits to the road wheels any turning movement imparted by the driver without reversing or carrying back to the operator the original movement of the road wheels.

Many attempts have been made to substitute another form of mechanism for steering gears; this consists of various rod, lever, chain, and spring combinations. All of these have failed, however, because they lacked the fundamental requisite of irreversibility.

Spur and Bevel Types. The spur- and bevel-toothed construction of gears may be reversible, and these types are to be found on low-priced cars, as the cost of cutting the gears is small. The spur gears have straight teeth, the edges, or sides, of the teeth being straight and parallel with the axis of the shaft on which the gear turns. In bevel gears the teeth taper toward a point and are inclined

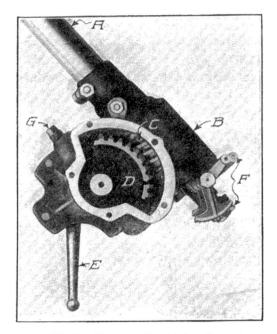

Fig. 8.   Worm and Partial Gear Type of
Steering Gear

to the axis of the shaft. Another construction is the spiral gear. Both types may be made reversible or irreversible as desired.

Worm-Gear Types. With a very few exceptions, automobile engineers favor the worm type of steering gear, and it will be found on the highest priced cars. It has the advantage of being reversible and is utilized in several forms. In the worm class of gears, some types are closely related, while others vary widely. For example, the complete sector and gear type differ only in that the wheel operated by the worm makes a complete circle or part of a circle. The full gear can be turned through 90 degrees and replaced on the shaft without presenting a new surface to the worm. Some hold that the worm must be subject to some wear, especially where it is most used. They contend that turning over the pinion brings new teeth to engage with the worm and that these teeth will not mesh properly when turned at an angle of from 20 to 30 degrees.

Worm and Partial Gear. Fig. 8 illustrates a gear of the worm and partial gear type. Advantages claimed for the design are durability, ease of action, and adjustability to wear. The parts are accurately cut and hardened, and the worm is provided with a ball thrust on either side. With this type, the teeth, which are in the middle of the sector and in mesh, perform the greatest work when the car is driven in a straight line and are most susceptible to wear. To compensate for this wear, the center teeth are cut on a slightly less pitch radius so that lost motion may be eliminated without affecting the upper and lower teeth of the sector and to prevent binding when turning at right angles. In the illustration, A is the steering column to which the worm C is secured, D is the sector in mesh with the worm, E is the ball arm, or lever, B the gear housing, F the spark and throttle bevel gears and levers, and G the lubricant plug.

Adjustment. Two principal adjustments are provided. End play of the worm is eliminated by loosening the jamb nuts and lock screws on the column housing. Displacing the oil plug G will disclose an adjusting collar which is set with a screwdriver. Adjust collar until all play is eliminated, but the worm must turn easily. The lock screws, above referred to, are so located in the gear housing that when one is directly over a slot in the adjusting collar the other is between two slots. Consequently, after adjusting the collar it is essential that the proper screw be selected for locking the adjustment. Both locking members must be prevented from turning, by using the nuts. Wear of the teeth of the worm and sector may be eliminated by means of an eccentric bushing, which, when turned, moves the sector into a closer relation with the worm. This is accomplished by removing a locking screw at the left of the ball arm and moving the arm, which turns the eccentric bushing. In case of extreme wear, it may be necessary to displace the ball arm and set the locking-screw section in a different position on the end of the hexagonal end of the eccentric bushing so as to bring the arm in such a position that it can be locked by the screw. End play of the sector shaft is eliminated by removing a locking arm and turning an adjusting screw in, after which the arm and lock screw are replaced and both set up tight.

Worm and Full Gear. A full gear and worm type of steering gear is shown in Fig. 9, with the gear cover removed. This type

Fig. 9. Typical Worm and Full Gear Steering Device

is irreversible, and the advantage claimed for it is that it can be easily removed and so readjusted that an unworn section of the gear may be brought into contact with the worm. This is a simple form, and it is possible to replace a worn gear with a new one, as the gears are not expensive.

Adjustment. The part most subject to wear is that section of the gear which meshes with the worm when the front wheels are traveling in approximately a straight line. Because of this wear, the teeth of the wheel are subject to deterioration. Usually the adjustment for the wear is made by bringing the worm into a closer relationship with the gear by using the eccentric bushings which support the worm shaft. This adjustment is practical when the lost motion is due to poor adjustment rather than to wear of the teeth. With the majority of types, it is possible to displace the steering arms, move the steering wheel about half a turn, then replace the worm wheel so that an unworn section opposite the worn teeth will be brought into engagement with a comparatively unworn portion of the worm proper. The eccentric bushings in this case can be utilized to obtain a correct meshing of the worm and gear teeth. End play of the worm can be removed by adjusting the ball thrust bearings on either side of the worm. Sometimes these bearings become dry, or the lubricant becomes gummy, causing the shaft to turn hard. Wear of plain bushings in the steering-gear case is responsible for lost motion; the remedy is to replace the bushings with new members.

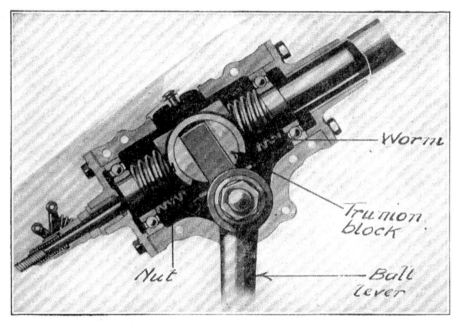

Fig. 10. Packard Steering Gear
Courtesy of Packard Motor Car Company

Worm and Nut. Next to the worm and gear, either full or partial, the form of steering gear most used is the worm and nut, which is made in several different combinations. Thus, the nut may operate the steering lever directly through the medium of a secondary lever, or it may actuate a block, which, in turn, moves either the lever direct or the secondary lever.

An excellent example of the worm and nut type of steering gear is shown in Fig. 10, which shows the Packard installation, used on

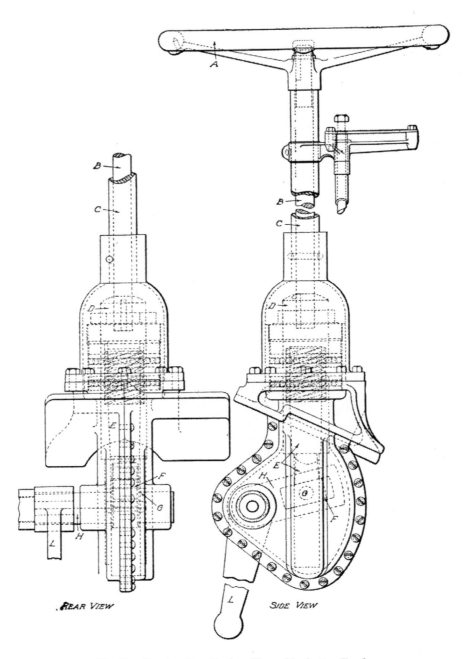

Fig. 11.   Steering Gear Used on Heavy Manhattan Trucks

the new "Straight-Eight" cylinder car. In this steering gear, the worm moves the nut which, in turn, moves the ball lever through a trunnion block which is held in a spherical seat. Attached to the ball lever are two projections which slide in the slots in the trunnion blocks.

In Fig. 11 is shown a form of worm and nut steering gear which is used on very heavy trucks and commercial cars. In this gear, the double worm is used; the inner worm carries, at its lower end, a block which is pivoted in a combination lever and shaft, to which the steering arm is attached. In the figure, A is the hand wheel turning the rod B within the steering-post tube C. This rod is driven into and keyed at its lower end to a member D which has internal worm threads. Another member E has a circular upper end on which are worm threads, while its lower end is slotted. The worm at the upper end meshes with the internal worm threads in piece D, while the lower slotted end carries, between the two arms of the slot, a rectangular block F. This block is hardened and ground all over and is fastened to the forked end of piece E by means of the hardened and ground pin G. This pin also passes through the arm H of the shaft to which the steering arm is attached. The steering arm is free to rotate about the center. This rotation moves the steering lever L in the arm of a circle.

The steering action is as follows: Turning the hand wheel turns the outer worm. This worm cannot move, so the inner worm is forced to move up or down, as the case may be, and moves the block with the pin through it, which, being fixed in the arm extension of the shaft, must turn the shaft. To this arm is attached the steering lever, so the latter must move. Although a rather complicated gear to explain and also to make, this gear, when finished, is an excellent one, and has been used for five or six years on heavy trucks with excellent results.

The Winton steering gear, Fig. 12, is not decidedly different from the one just shown, as will be noted by a close inspection of the parts. A is the internal worm, which is turned by the hand wheel, while engaging this worm are the block B and pin C, the block being partly cut away to show the engaging gear teeth. This block moves the jaw arm of the steering lever D. This jaw is not complete in this gear, but is cut away to save weight. The jaw arm, too, is connected directly with the steering lever, the jaw, arm, and shaft making one piece. The light work to which this was put made possible the economy in the number of pieces and in the weight of each. As before, turning the hand wheel turns the worm, which, in turn, moves the block and pin up and down and thus moves the jaw arm, which moves the steering lever.

Adjustment. The adjustment for lost motion in the worm and split-nut type of gear is generally made by loosening a cap screw on the column and screwing down an adjusting nut which has a right-hand thread. This adjusting nut acts directly on the thrust bearing, forcing the screw and half nuts, which slide, against the yoke rollers. In making the adjustment to a gear of this type, it is advisable to turn the road wheels to the extreme angle position, because the gear is the least worn at this point, and if it is adjusted only enough to take up the play when in this position, there will be danger of binding. Sometimes, when the adjustment is made with the road wheels straight, the gear will bind at the extreme positions.

Worm and Worm. In the worm and worm form of steering gear there is a worm within a worm, not wholly unlike the ones just described. Fig. 13 shows an example of this, which has a worm *C* attached to the steering

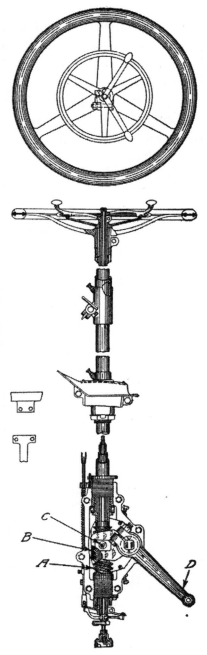

Fig. 12. Sectional Details of Steering Gear of Winton Cars
Winton Motor Car Company, Cleveland, Ohio

rod H, which is turned by the steering wheel A. Within and without this are worm threads, and external worm B meshing with the internal worm on the inside of C, while an internal worm D meshes with the external worm on C. The action of turning the hand wheel, then, moves one of these upward and the other downward.

The lower end B_1 of the inner worm member presses against a hardened end of the steering-lever arm E, while the lower end D_1 of the outer worm member presses against the other hardened end

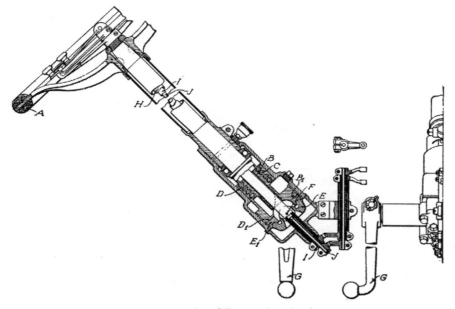

Fig. 13. Section of Gemmer Steering Gear
Courtesy of Gemmer Gear Company, Detroit, Michigan

E_1 of the same piece. There is no lost motion, or play, in the gear; when the hand wheel is turned, one worm rises and the other falls, as just described; the piece E will let one end rise and the other fall, as it is acted upon by the lower extremities of the two moving worms. This piece is pivoted at F and carries at its outer end the steering lever G, which thus moves in the customary manner. Within the steering post are the spark and throttle tube and rod I and J, which carry right through the whole gear and out at the bottom, where the spark and throttle-actuating levers are attached.

Adjustment. The adjustment of the worm and worm type, an example of which is illustrated in Fig. 13, is generally effected by a nut located at the upper end of the gear housing. This nut is provided with flats to accommodate a wrench hold. The end of the worm-wheel shaft is squared, and to this square the steering-lever arms are attached by means of a pinch clamp and bolt.

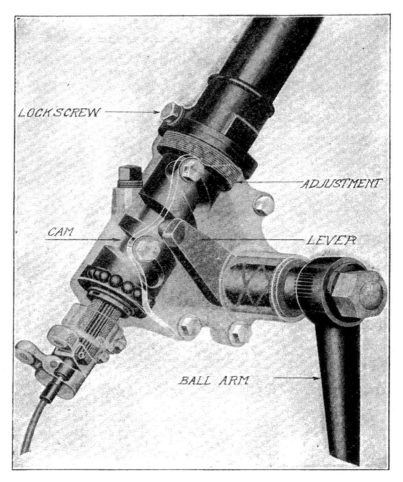

Fig. 14. Cam and Lever Steering Gear
Courtesy of Ross Gear Company

Worm and Gear. A variation of the worm type of steering gear is found in the Ross steering gear in which a worm, called a cam, is used to move a lever, Fig. 14. The worm or cam member is very similar to a single thread with variable pitch and is assembled between ball bearings which take both thrust and load. As the steering wheel is moved, the projection on the inside of the lever moves in the worm, or cam, up or down, and the steering lever is moved. The lever is attached to the shaft by a spline. The variable cam has the advantage for as the wheel is turned the movement of the lever

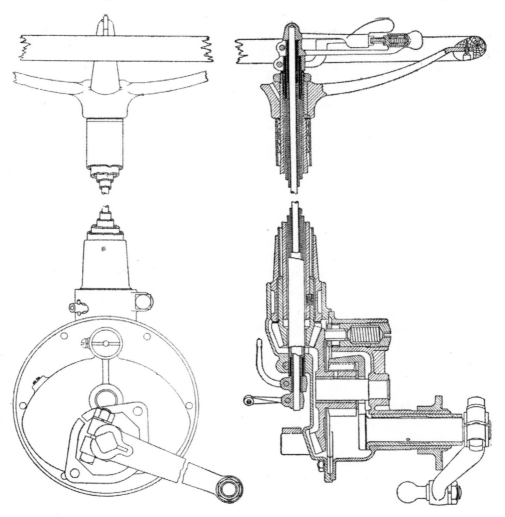

Fig. 15. Bevel Pinion and Sector Type of Steering Gear
Courtesy of Reo Motor Car Company, Lansing, Michigan

becomes more rapid at a greater turning angle. The variable pitch gives slow action at the straight-away position and practically eliminates the road shocks and gives the desired irreversible quality needed in steering gears. An adjustment for lost motion with up and down play is allowed by the introduction of shims under the thrust nut. To take up this lost motion, the lock screw should be loosened and the thrust nut backed off and one of the shims should be split and removed. This should be repeated until all lost motion has been removed but the adjustment should not be such that the steering gear turns hard or it will tend to strain the steering and most likely cause the balls in the bearings to break. For lubrication, the steering gear housing should be filled with a heavy liquid grease, such as 600 W.

Bevel Pinion and Sector. Among the other types of steering gears is that of the bevel pinion and sector, shown in Fig. 15. The bevel pinion moves the bevel-gear sector back and forth as it is turned, this motion being transferred to the steering arm attached on the same shaft to which the bevel sector is secured. This type of gear is said to be effective, but it is not irreversible, and shocks to the road wheels may be imparted to the steering wheel and move it.

Adjustment. The bevel and sector gear has two adjustments. The pinion may be moved up or down, as required, by unlocking the clamp bolts (one of which is shown at *D*) which permits the moving of the entire steering column up or down so as to obtain the proper relative position to the pinion and its sector. The position of the sector endwise may be adjusted by the block member *A*, which bears against a roller guide, forcing the sector into mesh more or less closely with the pinion. The spring *E* is provided to prevent rattling, and the screw *H* is a guide for the plunger and should not be disturbed in making the adjustment.

Hindley Worm Gear. There are a number of things about the Hindley type of worm which make it an excellent one to use for steering gears. A realization of this advantage is bringing about a greatly increased use of this form; so it will be appropriate and timely to look into its form, construction, and advantages.

The question of what makes the Hindley different from other worms naturally arises. The ordinary worm has the same diameter from one end to the other, the blank before the cutting of the teeth resembling a section of a cylinder. The Hindley, on the other hand, is not of uniform diameter, but has a smaller center diameter and enlarged ends. This gives it a waist, or hour-glass, shape.

An illustration will make this clear. Fig. 16 shows at *A* how the Hindley shape is generated and at *B* a finished gear, revealing plainly the reduced center diameter. In the upper figure, *EE* is the center line, or axis, of the worm, and *O* the center of the gear which is to mesh with it. *CD* is a circular arc struck from *O* as a center. If, on this curve *CD*, equal spaces be struck off, using a distance equal to the pitch of a single-threaded worm or the lead of a multiple-threaded one, as at *F*, and radial lines be drawn from the center *O*

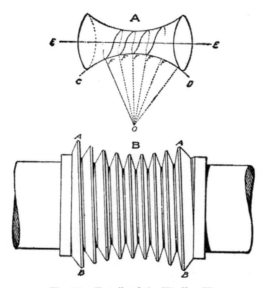

Fig. 16. Details of the Hindley Worm

to these points, these lines will be normal to the surface of the worm at those points; in short, the worm must pass through them, as roughly sketched in the figure. In the lower part *B*, of the figure, is illustrated a worm made on this principle, ready to be put into position.

This form of worm is used for the double reason of presenting more wearing surface—since it has at least three teeth in contact at any one time, as compared with one or at most two in the ordinary worm—and greater resistance to reversibility. The worm is used for steering gears because it is partly or wholly irreversible, its motion being a sliding one; nevertheless, all worms may be so cut as to be either wholly or not at all reversible. The sliding motion of the two parts in contact, as opposed to the rolling motion in the case of other mechanical movements of a similar nature, is greatly increased if there are three teeth in contact instead of the more usual one. If the friction of sliding be increased, the amount of reversibility will be decreased in the same proportion, for the added sliding friction will increase the natural reluctance of the worm to transmit power backwards. So much is this the case that it pays to use the Hindley form, despite its greatly increased cost of cutting.

Ford Steering Gear. The steering mechanism of the Ford car—a patented construction—differs radically from the conventional types in that its hand wheel does not directly rotate, or turn, the steering column or rod, but it imparts the necessary turning move-

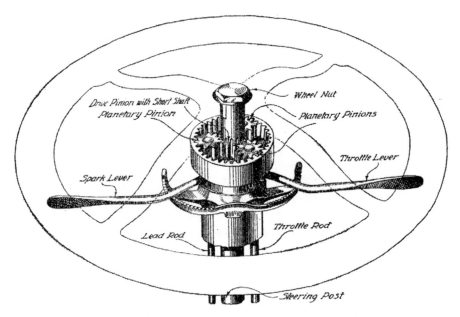

Fig. 17. Ford Planetary Steering Gear—An Unconventional Type

ment through the gearing and the use of a small shaft to which the hand wheel is attached. A phantom view of the gearing is shown in Fig. 17.

The steering column with its short shaft and drive pinion is enclosed in a tube or housing which is set at an angle and bolted to the dash. The housing does not extend the entire length of the column, as the lower end of it is mounted in a bracket that is rigidly bolted to the frame. The steering-gear post, or column, has a triangular flange at right angles to the rod, and each point of the flange has an integral stub, or pin, carrying a small spur pinion. The center of the rod is drilled and bushed to take a small shaft to which a fourth pinion, or drive pinion, is keyed. The upper part of the housing is shaped so as to provide a gear case, and the inner periphery of this case is cut to obtain spur teeth or, in other words, an internal ring gear. This gear is stationary.

The hand wheel is attached to the short shaft, and its drive pinion is held in place by a brass cover of the internal gear case. As the drive pinion of the shaft is in mesh with the three pinions mounted on the stubs of the steering column proper, and these three pinions are in mesh with the internal ring gear, any movement of the hand wheel will rotate the drive pinion on its shaft. This movement will cause the three spur pinions to rotate in an opposite direction against the internal gear, thus reducing the movement of the steering column as compared to that of the hand wheel. The three spur pinions compensate for any pressure of the drag link and the tie rod.

The operation of the Ford steering-gear mechanism explains the basic principle of the operation of the hand wheel; that is why the wheel is turned in the same direction that the driver desires the car to go. If the hand wheel is revolved from left to right, for example, the movement causes the three pinions mounted on the pins of the steering column to rotate from right to left; the pinions rotating against the stationary internal gear turn the steering rod in the same direction taken by the three pinions. The column swings an arm attached to it from right to left, and, as the rod is secured to this arm, it moves in the same direction, swinging the front road wheels so that they move from left to right, and to a degree that will correspond with the turning, or movement, of the hand wheel. It should be understood that the movement from left to right refers to the front half of the road wheels. If the driver desires to direct the vehicle to the left, the wheel is turned to the left.

The drag link of the Ford steering gear differs from conventional designs in that it is at right angles to the frame and is practically two-thirds the length of the tie rod. The end of the steering column is provided with an arm carrying a ball, and the drag link, or steering-gear connecting rod, as it is listed by Ford, has a ball-socket cap which fits over the ball of the steering rod. The drag link also has a ball socket at its other end, which fits over a ball arm on the tie rod. The tie rod, called the spindle connecting rod because it connects the spindles, is provided with yokes at either end, and these yokes are pivotally connected to the spindles by a bolt passing through them and through an eye in the spindle. The Ford drag link differs from others in usual practice in that it moves to the right and left, while those used on other cars move forward and backward. No provision is made with the Ford drag link for absorbing shocks or for automatically compensating for wear as usually is the case with the conventional type of drag link.

Semi=Reversible Gear. The steering gear used on commercial cars, particularly trucks ranging from 3- to 7-ton capacity, must not

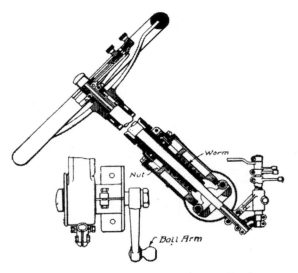

Fig. 18. Screw and Nut Gear Used on Trucks

only be capable of operation with a minimum effort, but it must absorb a great many of the minor shocks and a per cent of the larger shocks. The semi-irreversible type is most favored because of the above-named reasons. The design shown in Fig. 18 is of the screw and nut type. The nut is a solid piece, completely enveloping the screw, and the threads of the screw are in constant and complete engagement with the threads in the nut. The screw has a rotary motion and the nut has a longitudinal motion. The means of transmitting this longitudinal motion of the nut to the rotary motion of the steering arm is by circular discs at the lower end of the nut. These discs present constant bearing surfaces to the recesses in the nut, and are provided with slots into which the projecting levers from the rocker shaft fit. The screw pulls the nut up or down in the

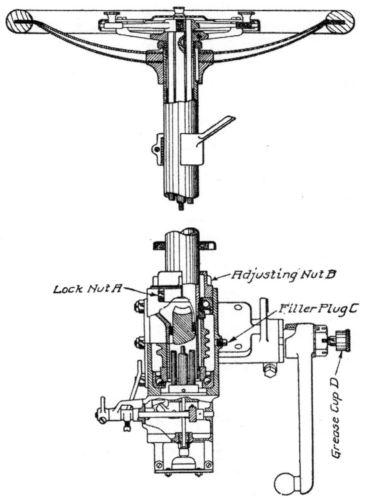

Fig. 19.   Worm and Gear Steering Arrangement—Semi-Reversible

housing, and there is no tendency for this nut to be moved sideways. The levers projecting from the rocker shaft into the swivels which rotate in the lower part of the nut are in direct line with the screw, so that the push and pull of the nut is in a straight line.

Removing Steering Gear. To disassemble the majority of steering gears it is necessary to remove the unit. With the type shown in Fig. 19, which is a semi-irreversible worm and gear, the removal may be accomplished by displacing the control levers at the top of the column and dropping the unit down through the frame. The adjustment of this type for end play is made by loosening the locking nut A and turning down the nut B until the play is eliminated.

STEERING WHEELS

Different Forms of Hand Wheels. *Wood Rim.* A variety of material is utilized in the construction of the wheel, which has super-seded the lever or tiller. The section or sections of the wheel or rim

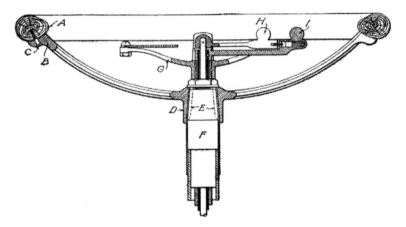

Fig. 20. Section through Typical Steering Wheel

are circular, oval, or elliptical; the oval, or ellipse, is turned upward. The strength of the wheel varies according to the material used and the process of assembly. The all wood wheel has not the strength of a built-up wheel with a metal core, but it is simpler and cheaper to manufacture. With the exception of the molded rubber type of rim, the majority of the wheels, particularly those fitted to high-grade cars, are built-up. Mahogany, circassian walnut, and black walnut are the materials favored. The wood is cut to short sectors of an annular ring of about 2 inches in width and so glued together as to eliminate joints.

The method of attaching the rim to the spokes of the wheel spider is by screws, and this method is illustrated in Fig. 20, *A* indicates the wood member, *B* the arms, or spokes, which have a boss through which the screw *C* passes into the wood. The hub of the spider *D* is attached to the steering post by two keys *E*.

Metal Core with Wood Covering. When the wheel design is made up of a metal core the ring is cast on the spider or integral with it. Coverings of wood concealing the ring are used, although with some types, a section of the ring may be noted. This type of wheel possesses great strength and the wood veneers can be secured at more frequent intervals than in the design previously described.

Different Wheels for Commercial Use. *Truck Types.* For the light delivery wagon, taxicab, and similar cars, no difference in the steering wheel is made, but when it comes to the heavier service, there is a need for a heavier wheel. This does not mean a heavier rim only, but a heavier, more rugged gear all the way through. The weight on the front wheels of a heavy truck is very great, and the tires, which are of solid rubber, may have frictional contact with the pavement of several inches in width. All this combines to make turning the vehicle from the driver's seat more difficult.

For this reason the driver must have a greater leverage, which means a larger diameter of the wheel. Then, too, the rim should be bigger in section in order to withstand the harder use of commercial service, and to provide for the large hands of the operators. Greater strain upon the rim of the wheel, on attempting to turn heavier weights with it, means that the rim must be fastened to the spider more securely. This means more arms, the four generally used for pleasure cars being increased to five for trucks. While this helps a great deal, since it provides five screws instead of four, it is not sufficient, and most of the big trucks today are equipped with steering wheels in which the rim is built over a central metal rim of the spider.

Pleasure-Car Types. Usual pleasure-car practice varies from 14-inch up to 16-inch wheels, while commercial car sizes begin at 16-inch and run up to 18-inch wheels on light trucks, and as high as 20- and 22-inch wheels on heavy trucks. Rim sizes vary considerably, a favorite for touring cars being an oval with from $\frac{3}{4}$- to $\frac{7}{8}$-inch vertical height and a length of about $1\frac{1}{16}$ to $1\frac{3}{16}$ inches. These figures have no connection with commercial work, the smallest being 1 inch and on up to $1\frac{1}{8}$ inches in height, with the long diameters varying from $1\frac{1}{4}$ up to $1\frac{3}{4}$ inches. For speed work, racing, and the like, it is usual practice for the operator to wind the surface of the wheel with string, this giving a rough surface upon which the hands will not slip. This is practiced, too, by many truck drivers, who claim that the strains of steering the big vehicle are felt less when the wheel is thus wound.

To preserve the nice appearance of the steering wheel and still give the roughened surface to which the hands will cling easily, even in wet weather, many manufacturers are making a wheel of knotted

Fig. 21. Knotted Wood Rim as Used on the Oldsmobile

wood, the use of this material allowing the formation of the wheel in any desired section, as is seen in Fig. 21. These wheels are usually made with a plain upper surface; the lower or under surface, however, being made in a series of depressions and humps, between which the fingers find a good resting place. This gives a good grip, as the under side of the wheel seldom gets wet.

Folding Steering Wheels. Although tilting steering wheels were introduced several years ago, they did not meet with favor until the Cadillac adopted them as standard equipment. The wheel, which is 18 inches in diameter and has an aluminum spider, is hinged to drop downward, a design facilitating entrance and exit at either side of the car and making it possible to attain the driver's seat without squeezing. The Cadillac wheel is shown in Fig. 22, while that used on the

Fig. 22. Hinge Type of Steering Wheel Used on Cadillac

King car, illustrated in Fig. 23, is of the tilting type. To operate the design, the wheel is turned until the wheel spider arm carrying the release button is convenient to the thumb of the right hand. The button is pushed to the right, and, by using both hands, the wheel is pushed forward and upward. The Herff type, shown in Fig. 24, is of the true hinged form; the rim is thrown up and out of the way, that is, the rim only, as the quadrant carrying the spark and throttle levers remains. There are several other types marketed, but their working principles are similar.

Throttle and Spark Levers. In the usual case, the arms of the steering wheel have the quadrant for the spark and throttle levers fastened to them. The levers are operated within the space inside

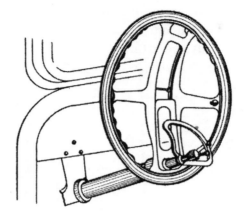

Fig. 23. Tilting Steering Wheel on the
King Car

of the rim of wood and above the spider of metal; the latter is usually at a lower level by several inches, as shown in the figure. In Fig. 20, however, the quadrant is not carried by the spider arms, but on a

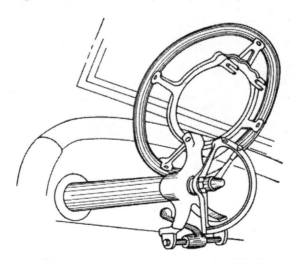

Fig. 24. Herff-Brooks Folding Steering Wheel

separate framework G, or spider of its own, up above the hub of the wheel. Over this framework the spark and throttle levers H and I work, serrations of teeth in the quadrant preventing the levers from moving, except when they are spring off by the pressure of the fingers operating them. In some cases, these teeth are done away with and friction surfaces are substituted; springs holding the contact surfaces together are so light as not to interfere with the moving of the levers by hand.

STEERING ROD, OR DRAG LINK

Operation. By the steering rod, or drag link, is meant the member connecting the ball arm, or lever, of the steering gear to the lever attached to the steering knuckle. This is clearly illustrated in Fig. 25. The steering gear is marked D, the steering arm pro-

Fig. 25. Typical Steering Arrangement on Pleasure Car

jecting down from it C, while the steering rod which connects the lower end of the arm with the lever on the knuckle is marked AB. F is the knuckle pivoted in the axle, which carries the two-end lever E, one arm of which has the steering rod attached to it at B, while the other carries the cross-connecting rod joining the two knuckles together. Since the pivot point is fixed, any movement imparted to the knuckle must result in its swinging about the pivot point and carrying the wheels with it.

This movement is imparted by the steering rod to the end *B* of the arm *E*. The steering rod itself simply connects with the steering lever *C*, swinging back and forth in a vertical plane with the steering knuckle *F*, which swings around and back in a horizontal plane, and imparts the movement of the lever to the knuckle. Since the end of the steering lever rises and falls and the end of the lever on the knuckle maintains a constant level, although moving in a circle, the rod must have a universal joint at one end. This is really a necessity from two points of view: to allow the rear end to move up and down vertically while the front end swings around

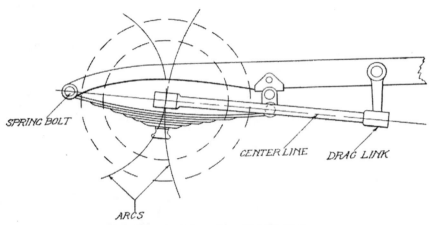

Fig. 26. Correct Drag Link Position

in a circle; and also to allow the front end to swing in a circle set in one horizontal plane, while the rear end remains stationary or practically so in that plane. In short, the two ends move continuously, each in its own plane, but the two planes never coincide—the one is always vertical, while the other always stays horizontal. This necessitates at least one universal joint. Many makers play on the safe side, and lower the cost of production by making the two ends alike—a universal joint on each one.

The ball arm should be of such length that the rear end of the drag link is always lower than the front end. To get the best possible ease in steering, the drag link should never be parallel to the road surface. To check this condition, a line drawn through the center of the drag link should also pass through the center of the front spring bolt. Fig. 26 shows how the length of the ball arm

160

has a certain control over the position of the drag link. As the wheels meet the deflection of the road, there is a point around which the axle tends to revolve and this point is the front spring bolt. The drag link also tends to revolve around the end of the ball arm. If these two points of revolution be considered, it will be seen that the axle will form an arc around the front point or spring bolt, and the drag link will form an arc around the ball arm. These two arcs should be tangent to each other as shown in the figure. The longer the drag link, the less variation there will be between the two arcs. It will be readily seen that if the ball arm is shortened, the center line of the drag link will not pass through the center of the spring bolt and the steering will not be as easy as it should be.

Types of Construction. A glance at the construction shown in Fig. 27, and also in Fig. 28, shows a steering lever made with a ball end, or partial ball end, upon which the steering rod is hung. In this, the partial ball is formed in the center of a bar, the inner end of which is threaded and screwed into the steering arm with a nut on the outside to prevent its backing out. The ball itself is made separately and slid on over the rounded end of the shaft, or axis. After this a sleeve is put on, followed by a nut which holds the sleeve up tight against the ball. The function of the sleeve is to give the spherical end of the rod plenty of play in a sidewise direction.

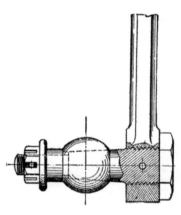

Fig. 27. Steering Lever with Ball End

This is a cheap form of construction, but could have been made in one piece had it been desirable or necessary to do so. Such a form has a metal-to-metal contact, which is hard upon both ball and socket, necessitating frequent and costly replacements. These replacements are obviated by backing the ball socket up with a spring or springs, as is shown in Fig. 28. This form of construction is now quite generally used: the socket of the ball in the inner end of the rod is set inside of a sleeve with a spring on each side of it. These springs not only take up the road shocks but the wear as well, the shoulder against which they rest being adjustable. In this figure, J is the lower end of the steering lever with the ball end. This lever is mounted in the ball socket G. A is the body of the steering rod, which is expanded at the end to a larger diameter, this being designated in the figure as B. Within this expanded portion, the sleeve E at one end acts as a shoulder for the spring F.

At the other end, the outside of the sleeve is threaded to receive the collar C with the hexagon end K. Within this, a second spring

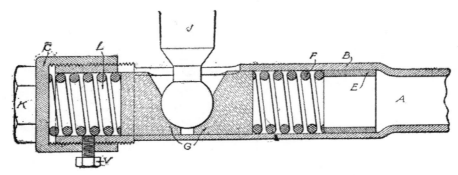

Fig. 28. Adjustable Form of Ball-End Steering Rod

L holds the socket up to its position. The location of the collar C determines the tension of the spring L, and this is locked in its position by the screw V. Should there be wear, which necessitates

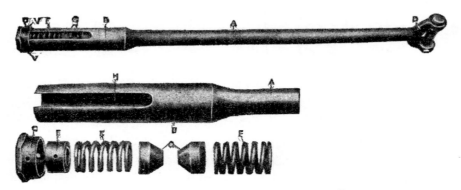

Fig. 29. Cross-Connecting Rod Assembled and in Parts

the moving of the ball toward the open, or left, end, the whole thing is disassembled and a longer sleeve inserted in place of the one shown at E. On the other hand, ordinary wear is compensated for by taking up on the collar C, first loosening the lock screw V.

In Fig. 29, a rod is shown assembled at the top and disassembled into its components at the bottom. The two ends differ, one being but a simple yoke with a plain bolt through it, marked *D*. The other, however, is a ball end with an adjustment and with springs to take up shocks.

All these parts are marked in the figure and may be located by letter. The body of the rod is marked *A*, the expanded end *B*, which has a groove *H* cut in it. Into the inner end of this groove is fitted, first, the spring *F;* second, the two halves of the ball socket *G;* and third, another spring. The sleeve *E* closes the outer end, and over the exterior is screwed the adjusting nut *C*. The nut and sleeve are held in place by the locking pin *V*, which passes through the outer nut, the shell end of the rod, and the inner spacing sleeve, the ends being riveted over to hold it in place. This form limits the adjustment to a full half turn of the nut, while the pin would soon need replacement if much adjusting were done, as some of its length would be lost each time it was riveted over because of the chipping away to allow it to be taken out.

Cross=Connecting, or Tie, Rods. The object of the cross-connecting, or tie, rod is to connect the right- and left-hand steering knuckles so that the road wheels will be turned alike. The general practice is to place the rod back of the front axle, a location avoiding the possibility of damage if an obstruction in the road is encountered, but in some instances the tie rod is placed in front, as in Fig. 25.

The tie rod is made adjustable to compensate for any change that may be necessary to preserve the alignment of the wheels, and, generally, the rod is adjustable at either end. The yoke ends of the rods are made adjustable, screwing on or into the rod proper and secured by lock nuts or other suitable fasteners. The adjustment is easily made. Decreasing the length of the rod increases the gather, or distance, between the forward section of the front wheels, while increasing it causes the wheels to toe in. This applies to the tie rod behind the axle.

Gather. Gather, or tow-in, is the setting of the front wheels so that the distance between the front of the wheels is less than the distance at the back. This distance is measured between the wooden felloe of the wheel at a height equal to the center of the axle measured from the ground. To do this correctly, the wheels should be jacked clear of the floor and the distance measured with a stick, with the valves of the tires in one position, which is on a level with each other, at the felloe of the wheels. After getting this distance, the valves should be moved to the back of the front axle and the distance measured again between the felloes. There should be a distance of $\frac{1}{4}$ inch between the two positions. Fig. 30 shows how to measure for gather.

Function and Shape of Steering Knuckles. The steering knuckles serve as a pivot for the road wheels, enabling them to move in a

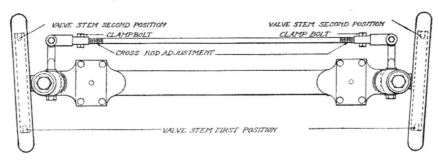

Fig. 30. Measuring for Gather

horizontal plane. The design of the knuckle depends upon the axle, and the pair used on a car are different as one has a lever for carrying the drag link. Both have integral spindles to which the tie rod is attached. Figs. 31 and 32 illustrate the difference between the knuckles. Fig. 31 shows a right knuckle, forged from a blank of chrome nickel steel, while the one at its side is the finished part. *A* is the place for the outer wheel bearing, *B* the position of the inner bearing, *C* the hole for the pivot, or knuckle, pin, *D* the upturned steering arm, and *E* the arm to which the tie rod is attached. Fig 32 is an example of a left steering knuckle of the same pair, both before and after machining. The letters in Fig. 31 apply to this knuckle.

Lost Motion and Backlash. Lost motion of the steering wheel does not always indicate that the steering gear is at fault, for wear in the steering-gear assembly usually takes place first in the clevis pins, yokes, and connections of the drag link. The spindles, spindle bolts, and wheel bearings are factors. Despite the fact that the front road wheels are deflected but a few degrees the spindles, bolts, or bushings may be worn, as these parts are subject to radial and thrust loads.

The spindle bolt, which does not move, tends to wear oval; adding to this tendency the wear of the spindle bushings, one has considerable lost motion to contend with. Wear of the wheel bearings contributes to the apparent lost motion of the steering gear as do the connections of the drag link. Taking all of these factors into consideration, and allowing but a small fraction of an inch for play of each worn part, the sum total may result in considerable movement of the hand wheel before the road wheels are deflected.

Fig. 31.   Finished S.G.V. Chrome Nickel-Steel Steering Knuckle and the Same before Machining

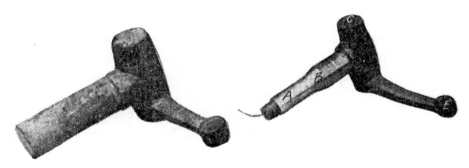

Fig. 32.   Left Steering Knuckle of S.G.V. Car before and after Machining

Lost Motion in Wheel. While there should be a certain amount of movement to the hand wheel before it actuates the road wheels, the lost motion, as a rule, does not exceed $\frac{1}{4}$ or $\frac{3}{8}$ inch when the gear is new. This amount is essential as without some free movement the steering of the vehicle would be tiresome. Wheels may be keyed or pinned to the column. When play exists as the result of a worn key, pin, or slots, the remedy is to re-cut the seats and make and fit a new key or pin. With some types of wheels the use of a wheel puller will be necessary to displace them. Another cause of lost motion, when the wheel is tight and linkage free from play, is a loose key retaining the worm or gears of the steering gear proper. A simple test of the hand wheel is to hold the tube, or post, securely and move the hand wheel. The amount of play in the drag link can be ascertained by grasping it about midway and trying to move it backward or forward or in the normal direction of travel. Hold the ball arm of the steering gear when making this test.

Lost motion in the drag link is caused by the ball and the blocks that rub against the ball becoming worn. In practically all drag links, there is an adjustment to take care of this and it is to be found at both ends of the drag link. In some drag links, the adjustment consists of a cap that screws over the end of the drag link tube, while in other types it is a slotted thread piece that screws into the tube. The spring tension is increased when the adjustment is tightened and holds the block in closer contact with the ball, thus removing the lost motion. The adjustment, ball, blocks, and springs are shown in Fig. 28.

The amount of backlash present in the irreversible and semi-irreversible types of steering gears may be determined by disconnecting the drag link, grasping the ball arm, and moving it up and down and back and forth. Worn bushings in the steering-gear case are frequently the cause of movement of the column as a whole. Another component that should not be overlooked in the search for the cause of lost motion is the ball arm. Movement of this member on its shaft can usually be eliminated by tightening the nut.

Wheel Wabble. This trouble may be caused by several things or a combination of things. Loose steering gear parts, such as steering knuckles and steering-knuckle pins, axle out-of-line, axle needs more caster, or too much backlash in the steering wheel. A careful inspection should be made for loose parts and if any parts show wear they should be renewed. Tighten all loose parts so that the steering gear is a little tighter than before. The axle should be lined up in regard to gather, camber, and caster, and the wheel bearings should be in good condition and correctly adjusted.

Cross-Rod Connections. Lost motion, noise, and rattle can often be traced to loose cross-rod connections. This part of the steering gear should be inspected occasionally because the clamp bolts, that hold the connection tight, are very apt to work loose. This not only causes the wheels to get out of alignment but it is also very dangerous as the thread on the cross-rod might strip and all steering control would be lost resulting in a serious accident. If

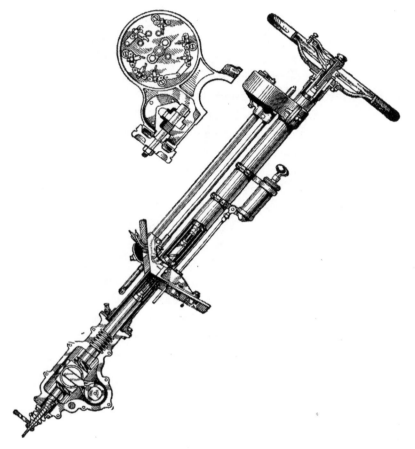

Fig. 33. Packard Steering Gear Parts

the connections are allowed to remain loose, the rod and connecting clevis will become worn with the consequent need of renewal of all parts concerned.

Tires Wearing Out Quickly or Unevenly. Although the tires are not actually a part of the steering gear, the condition of the steering gear has a great deal to do with the amount of service obtained from the tires on the front wheels. If the wheels are not correctly lined up, that is, have the correct amount of gather and camber, the front-wheel tires will wear out very quickly. Tires have been known to wear out within a few hundred miles from the above cause. When driving toward the curb, care should be taken not to strike the curb because this shock will often throw the wheels out-of-line and the driver will not notice it and the damage to the tires will occur before the condition is discovered. Worn steering-knuckle pins and bushings will alter the camber of a wheel and will cause tire wear. A rim that is not true on the wheel, as well as a loose rim, will cause excessive wear on the tires.

Lubrication of Steering=Gear Assembly. The proper lubrication of the steering-gear assembly adds to its life, but this work is not, as a rule, thorough. The steering gear proper should be packed with grease, the ball and socket joints of the drag link and steering-arm lever with a light grease; the clevis pins also should be lubricated. The steering-knuckle pins are provided with either grease or oil cups.

A point generally overlooked in the lubrication of the steering gear is the steering-post spark shaft and throttle-sector anchor tube, shown in the illustration at Fig. 33, which is of interest in that it illustrates the assembly of the Packard car. The post carries the control-box unit. The spark shaft and throttle tube frequently lack lubricant and should be cleaned and coated with a graphite grease before replacing when the gear is being reassembled. The lower extremity of the spark and throttle members carry levers or small bevel sectors which operate the linkage of the ignition apparatus and carburetor. Clamping screws are generally used to secure these parts.

SIDE VIEW MOON FOUR-DOOR DE LUXE SEDAN FOR 1926

FRAMES AND SPECIAL TYPES OF DRIVE

FRAMES

Characteristics of Parts. *Frames.* The chassis frame practically is the foundation of a motor vehicle, since all of the power transmitting and other units are attached to it. Motor-vehicle construction depends, to a certain extent, upon the general design of the chassis, the construction of the power plant and transmitting units, their mounting, the method of final drive, the wheelbase, etc. The size of the material used depends upon the weight of the units carried and the capacity of the vehicle, and varies from thin and small sizes on very light pleasure cars to heavy structural I-beam frames on commercial vehicles.

The use of pressed steel is becoming more popular, as is also the tendency to narrow the frame at the front to obtain a shorter turning radius. Designers favor a kick-up at the rear, which affords better spring action and permits of a low suspension of the body.

General Characteristics. When the automobile was first introduced, comparatively little attention was paid to the frame, as the other components of the chassis, such as the power plant, gearset, axles, etc., were held to be of greater importance, consequently the frame did not receive the consideration it should. The first types of automobile frames were made of wood, ash being the material used, the cross members, used for the different unit supports, being forged out of angle iron and bolted to the main frame.

For light car construction steel tubing was used. The supporting brackets being bolted on the main frame, and in some cases they were brazed to and made a part of the main frame. After experiencing considerable difficulty, however, due to accidents and other failures which were traced directly to poor frame design, the automobile engineer found that it was possible to build a frame of great strength with less weight than the troublesome types.

The improvement in frame design is the result of the tendency to provide perfect alignment of the power plant, clutch, and gearset, making use of what is known as the unit power plant on some models, while on others, particularly of the heavier type, flexible mounting of the units has been resorted to. The tendency is toward the use of a flexible mounting of all individual units, at least to some degree, in order to relieve them of the stresses brought about by frame weaving when the road wheels mount an obstacle on the road surface.

Classes of Frames. The most prominent types of frames, divided according to their use, are the pressed-steel frame, the structural frame, and the structural I-beam frame; the latter is confined to commercial cars. These classes may be subdivided according to the general construction and material, as well as to the distribution of the chassis units.

The material employed is either pressed or rolled steel. The wood frame or combinations of wood and metal frames are practically a thing of the past, and are to be found, with one or two exceptions, on old cars. The steel frame may be constructed in the following shapes: channel, L-beam, angle, T, Z, tubing, flat plates, and combinations of any two or more of these. Other forms are possible. For example, the channel may be turned with the open side in or out, the two constructions being widely different; or the angle may have the corner down and out, down and in, up and out, or up and in. Similarly, the T-shape may be a solid T turned up or down, or it may be a hollow T-section with space between what might be called the two sides of the leg; this shape may be turned either up or down, while the Z-shape may be turned horizontally or vertically. Many frames are constructed with the open end of the channel section turned in, and use is made of a steel underpan of flat section attached to the under side of the main frame. In several instances there is a tendency to make the frame and underpan as one piece, in which case the frame section assumes the shape of a channel with an exceedingly long lower flange.

Another type of frame is that having a continuous section throughout. Others have a varying section. Thus, the ordinary steel frame of modified channel section may have a depth of perhaps 5 inches at the center, a width of upper flange of $1\frac{1}{4}$ inches, and a width of lower flange of 2 inches. A frame similar to this would taper down to the ends to perhaps 20 inches in vertical height, and to 1 inch in width of both top and bottom flanges. Then, again, frames which are bent upward or downward at the ends or in the middle really differ from those frames which preserve one level from end to end. The practice of bending the chassis frame is very prevalent of late, the upturning of the ends bringing about a lower center of gravity, making for stability and ease of entrance and exit to the body.

Tendency in Design. There is a marked tendency toward making the chassis frame wider at the rear and narrower at the front. In one or two cases the designer appears to have gone to the extreme in this respect. The advantage of the narrow front construction is that it enables the car to be turned in a shorter radius. The use of a wide rear frame provides more space to support a wider body. A more recent development is to make the longitudinal bars of the frame parallel over the front spring and near the rear spring, and to have them tapered from behind the front to the rear springs. A certain amount of material is said to be gained by this construction, as no heavy reinforcement or sudden offset is necessary to the frame. By

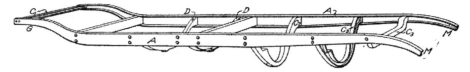

Fig. 1. Typical Automobile Frame of Pressed Steel

widening the frame at the rear it makes possible the placing of the springs directly underneath the frame. Some car makers have the sides of the frame straight over the entire length, but tapered from the front to the rear.

Fig. 1 illustrates what is termed a single drop or a kick-up. This is a type of pressed-steel construction, of channel section, and the deepest and strongest section is at the center where the greatest stresses occur. Some frames are built with a double drop, having a downward bend just forward of the entrance to the rear part of the car body, followed by an upward turn just back of the same entrance. The upward turn at the back is carried higher than the main part of the frame for the purpose of obtaining a low center of gravity. Then there is what is termed the bottle-neck construction, a bend inward which resembles that in the neck of a bottle. This obtains a short turning radius. Originally, frames were narrowed in front, the difference in the width between the front and rear being at first an inch or so on each side, gradually increasing until it became 5 and 6 inches. This type did not prove efficient, and the trend favored the taper previously explained.

A not uncommon form of frame is shown in Fig. 2, which compensates for an abnormal rise of the rear axle without the possibility of its striking the frame. Some frames have a bend at the ends to take the spring fastenings.

Pressed=Steel Frames. The pressed-steel type of frame is very popular with designers and is largely used on commercial cars up to and including 1-ton capacity. This is popular because it is the lightest in weight for equal strength of the structural iron or rolled channel and I-beam section. The cost of pressed steel is somewhat higher,

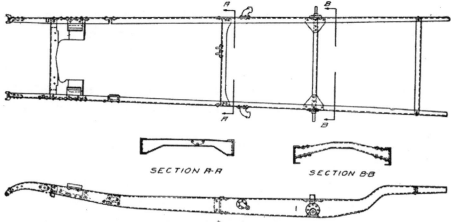

Fig. 2. Frame of Sterns-Knight Car in Plan

because it is heat-treated material used to obtain maximum strength. The cost varies with the section, material, and the nature and extent of bending. The finished frames are easy to handle, and the assembling cost is small. The channel shape is easy to brace and repair. These and other advantages have brought about its use.

The cheapest construction is the straight side rail, and, when conditions permit, it is usually tapered at front and the rear, and the forward end is sometimes shaped to receive the spring hangers. When the side members are inswept to permit a short turning radius, it is necessary to make the flanges of the side rail of considerable width at this point, tapering gradually to the rear, to provide the proper strength at the point of offset.

Sub=Frames. The modern tendency is to eliminate the sub-frame—a step due to the flexible mounting of the power plant and unit construction—because it simplifies the frame. It has also been made easier by the tapered frame, which is narrowest at the front where the units are attached. The most common method of supporting the engine is the three-point. Sub-frames are used, however, as they serve the purpose both of supporting some unit and of strengthening the frame.

Sub-frames may be of two kinds, viz, those in which the sub-frame is made different for each unit to be supported, and others in which one sub-frame supports all units regardless of size, shape, or character of work. The type of sub-frame made to support each unit usually works out to two pairs of cross-members, one for the front of the unit and one for the rear; while the type which supports all units regardless of size works out to longitudinal members, supported, in turn, by two cross-members, front and rear. The added weight for the first-mentioned type is less than for the other, since it comprises only four cross-members; while the last-named type consists of two cross-members equal to two of the others and of two very long members parallel to the main-frame members, each much longer and thus much heavier than the corresponding cross-members. In the two frames already shown, Fig. 1 shows the unit type of sub-frame with only cross-members, while Fig. 2 shows the more modern type in which the power plant is of the unit type and rests directly upon the main frame, being the three-point suspension type in which the forward point is on a frame or special cross-member, while the rear two points are the crankcase supporting arms resting directly on the main frame.

Rigid Frame. A pressed-steel or rolled-stock rigid frame has its advantages, particularly with reference to the commercial vehicle. It permits the body to be rigidly secured to it, and as it does not give with the inequalities of the road, the body is not racked. An advantage of the rolled stock is its cheapness, except, of course, for the lighter models of the assembled type for which frames can be secured at low figures. Another advantage of the rolled stock is the ease with which the wheel base may be altered.

Effect on Springs. The effect of frame construction upon the design and duty of springs should be considered. This feature

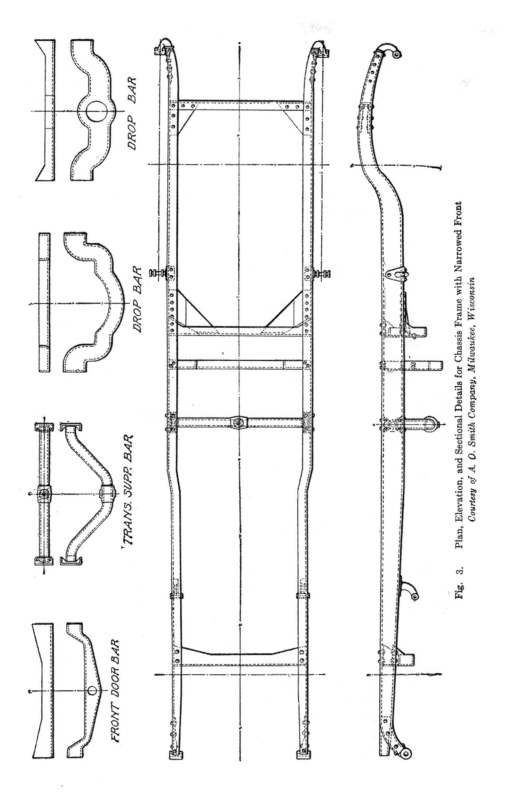

DROP BAR

DROP BAR

TRANS. SUPP. BAR

FRONT DOOR BAR

Fig. 3. Plan, Elevation, and Sectional Details for Chassis Frame with Narrowed Front
Courtesy of A. O. Smith Company, Milwaukee, Wisconsin

is not generally understood, but it has an important bearing upon the life of the car. A rigid frame relies upon the springs to allow for all axle displacement. If a front and a rear wheel on opposite sides are raised several inches at the same time, the frame is subjected to a torsional stress. If the frame is rigid, springs of considerable camber must be employed in order to absorb the shock without being bent past the limit of safety, and they must be sufficiently flexible to absorb all the shock without any tendency to lift the other wheels from the ground. To accomplish this shock absorption, a different type of spring is used on a rigid chassis from that employed on a flexible frame. The use of underslung spring suspension has come into favor for this reason, as it permits the frame to be carried fairly low, without sacrificing spring camber or necessitating a dropped rear axle.

The flexible frame, when diagonally opposed wheels are raised, does not impose all the stresses on the springs but it absorbs a part of them. For this reason, springs on a flexible chassis are flat or nearly so, with a limited amount of play. Flexible construction also permits the frame to be carried equally as low as with the underslung spring, and yet the spring is perched above the axle, where it is more nearly in line with the center of gravity, thus reducing side sway.

TYPES OF FRAMES

Pressed Steel. Pressed steel is purchased in sheet form, cut to the proper shape in the flat, and then pressed into channel form under great pressure. It is made of steel rolled into sheets and is somewhat closer grained than ordinary steel. There is no breaking of the flake in the rolling process. The pressed-steel frame, as previously pointed out, permits of greater simplicity in assembling, since the parts can be easily bolted or riveted. Fig. 3 is of the type of pressed-steel frame having a tapering section, a kick-up at the rear end, five cross-members —one of them a tube—and is narrowed in at the front to give the largest steering lock. Otherwise it presents only standard practice.

Wood. Wood is universal and easy to obtain. While no longer classed as cheap, it is not expensive; moreover, wood is kept in stock nearly everywhere. Users of wood for side-frame members claim that the wood frame is not only lighter but stronger. In addition, the wood frame would undoubtedly possess more natural spring and

TABLE I

Comparative Strength of Steel Channels and Laminated Wood Frames

Material	Size (in.)	Weight per Linear Inch (lb.)	Resisting Moment	Resisting Moment per Unit Weight
Pressed Steel	$\frac{3}{16} \times 4\frac{1}{2} \times 1\frac{3}{4}$	.408	114,830	280,955
Ash	$1\frac{3}{4} \times 6$	.266	142,275	534,870

resiliency, so that it would make a lighter and easier riding frame. A section of a wood frame is shown in Fig. 4.

This shows a frame made of laminated wood. There are three very thin sections of selected ash, marked A, which are glued together,

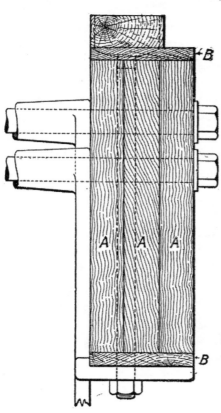

then screwed and bolted to prevent the glue from opening up. To further this purpose, a strip B is fastened on the top and bottom in the same manner. These strips are laid with the grain running horizontally, while the main pieces are laid with the grain running vertically. This construction makes a very strong and light-weight frame; the comparative figures for a steel section and the section shown, as given from the tests of the engineers of the Franklin Company, is shown in Table I. These tests, which are authentic, seem to bear out the contention that the wood frame is both lighter and stronger than the steel frame. Some wood frames have a thin steel plate put inside the frame which helps to make it stronger.

Fig. 4.   Section through Wood Side-Frame of Franklin Car

Recent Types of Frames. An innovation in frame design is the Marmon, shown in Fig. 5, the side rails and running boards of which are made in a single unit. The great width of the running board, varying from $11\frac{1}{2}$ to 16 inches, serves as the bottom flange of the frame, and is therefore of Z-section. The vertical section of the frame is 10 inches high, and has height enough to replace the running board fenders without appearing narrow. At the front and rear ends of the frame, the running boards are curved upward, strengthening the frame

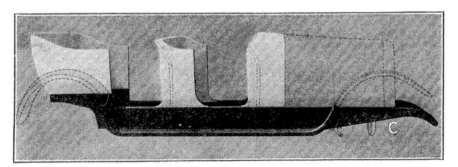

Fig. 5. Marmon Aluminum Frame, Showing Running Board Construction

as well as supporting the fenders into which they merge. The frame, beyond these points, both forward and rearward, is made of channel section of the conventional type. The rear of the frame is 45 inches wide and tapers to 30 inches at the front spring hangers. The great depth of the frame section makes it very stiff, so that the body sills

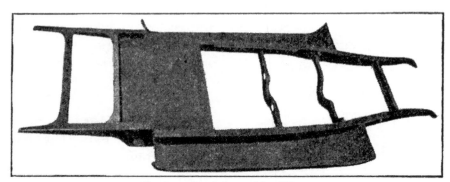

Fig. 6. Brush Pressed-Steel Frame
Courtesy of Hale and Kilburn Company, Philadelphia, Pennsylvania

can be entirely eliminated, and yet the doors will not work loose or bind when the top is up or down.

Fig. 6 illustrates a type of frame similar to the Marmon, the Brush frame, controlled by the Hale and Kilburn Company, of Philadelphia.

Steel Underpans. The underpan has assumed a great deal of importance in the last two years, for makers have more and more realized that it is highly important to protect many of the parts from road dirt, flying stones, water, etc. Designers have, therefore, given

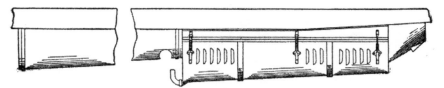

Fig. 7.   Two-Piece Pressed-Steel Underpan Used on Winton Cars
Courtesy of Winton Motor Car Company, Cleveland, Ohio

considerable attention to its shape, size, and method of attachment. In some types, it apparently runs underneath both engine and transmission and is made more or less a part of the main frame. Therefore, its quick removal on the road would be difficult, if not impossible; yet road accidents sometimes make it necessary for the driver to take this pan off to get at the lower side of engine, clutch, or gear box.

For this reason, underpans generally resemble more closely that shown in Fig. 7. This is a side view, showing the semicircular form of the pans, as well as the two-piece construction. The forward part under the engine, which would be taken down fairly often, is held in place by three spring clips on either side. Lifting these clips off is only a second's work; in addition, there is a filler piece in front, helping to make the pan fairly air-tight. The depth of the pan increases slightly toward the rear, so as to form a slope down which liquids will drain; the rear end is fitted with an upturned elbow, so that it will not drip until it accumulates a considerable quantity of liquid. Continual dripping indicates a full charge, and the pan is drained by turning the elbow over.

Fig. 8.   Detail of Spring and Section of Winton Underpan

Fig. 8 is a detail of the arrangement of the pan shown in Fig. 7. This indicates both the permanent part of the underpan, which is attached to the frame, and the removable part, which is freed by loosening the spring clips shown.

Rear=End Changes. The locating of the fuel tank at the rear of the chassis—a practice that was brought into favor largely through the introduction of the vacuum system of fuel supply—has resulted in a number of changes to the rear ends of frames. The placing of the fuel tank at the rear is not new, and probably it would not have occasioned any change to the rear end of the frame were it not largely for the fact that the spare tires are now carried at the rear of the

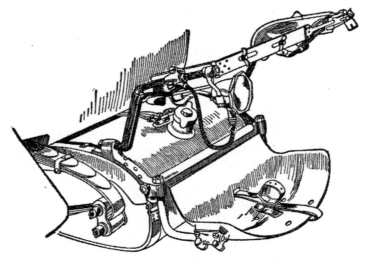

Fig. 9. Solid Rear Construction of Locomobile for Tires and Tanks
Courtesy of Locomobile Company of America, Bridgeport, Connecticut

chassis. The tires themselves are not heavy enough to make it essential to strengthen the rear ends, but the very general use of carrying the spares inflated on demountable rims has added considerable weight to the rear of the chassis. This weight, coupled with that of a large fuel tank, has compelled makers to give more attention to the rear construction.

Provision is made for carrying the spare tires on the Locomobile chassis by means of an apron conforming in shape to the shoe. The three-quarter elliptic springs of the scroll type have ends attached to the outside of the main frame, which is carried back and serves as an extension for attaching the fuel tank. A cross-member is also utilized; it serves as a point of attachment for the two rods supporting the lower apron and for two upper rods as well. This design has merits in that the tire carrier is firmly anchored and serves to protect the fuel tank from injury possible in operating in crowded traffic where rear-end collisions are not uncommon. As may be noted, Fig. 9

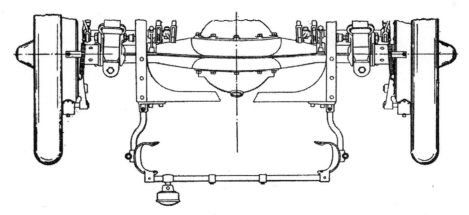

Fig. 10.   Sketch of Rear-End Construction of Reo Car

shows the method of using an upper cross-member to prevent theft of the tires.

A different type of rear construction is shown in Fig. 10, a Reo. Here the rear cross-member is gusseted, and a pair of substantial arms

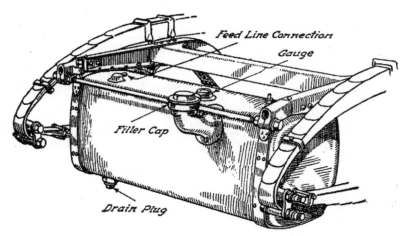

Fig. 11.   Typical Rear-End Construction, Carrying Gasoline Tank

are riveted to the cross-member. These arms serves as an anchorage for the tire holders which, in turn, have a cross-rod for protection. Still another design is shown in Fig. 11. Here the side rail of the frame projects back of the rear cross-member of the frame for a distance of about 12 inches. The fuel tank is suspended from these two extended frame members by means of steel straps which pass around the tank.

Commercial=Vehicle Construction. Commercial work, being rougher, harder, and cheaper work, changes the frame construction just as it does everything else about the car. In Fig. 12, a commercial-vehicle frame which brings out this point is shown. The main sills are 6-inch channels, while all of the other members are correspondingly large angles and channels. In one place the section consist of a box shape made up by bolting two large channels together, with the open sides in. The total overall length is not given, since this differs according to the variations in the wheel base; but, by a comparison of the figures given, it is seen that the frame shown is in excess of 210 inches long by about 37 inches outside width. This is about twice the total length of the average small car.

In the bracing and arrangement of the different members, this frame shows other points of difference, the cross-members, for instance, being nine in number, not including the two diagonal cross-members. The longitudinal members, too, are eight in number, not counting the two diagonals.

FRAME TROUBLES AND REPAIRS

The more usual troubles which the repair man will encounter are sagging in the middle; fracture in the middle at some heavily loaded point or at some unusually large hole or series of holes; twisting or other distortion due to accidents; bending or fracture of a sub-frame or cross-member; bending or fracture at a point where the frame is turned sharply inward, outward, upward, or downward.

Sagging. A frame sags in the middle for one of two reasons, either the original frame was not strong enough to sustain the load or the frame was strong enough normally, but an abnormal load was carried, which broke it down. Sometimes a frame which was large enough originally and which has not been overloaded will fail through crystallization or, in more common terms, fatigue of the steel. This occurs so seldom, and then only on very old frames, that it cannot be classed as a "usual" trouble; moreover, it cannot be fixed. .

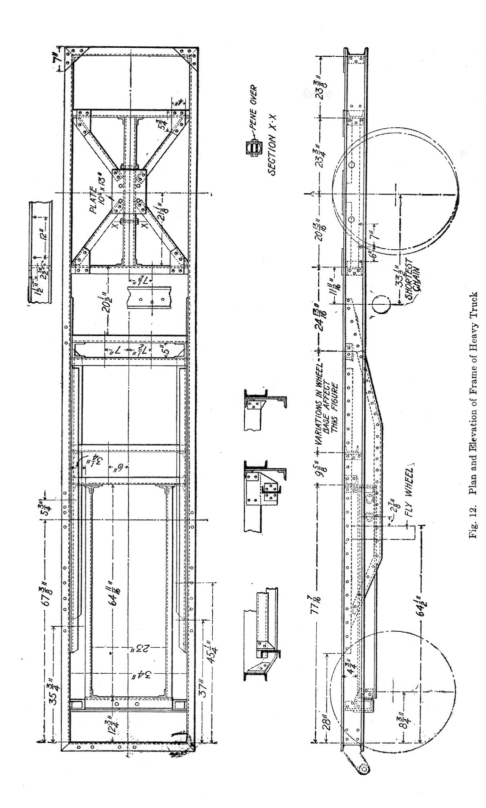

Fig. 12. Plan and Elevation of Frame of Heavy Truck

183

When a frame sags in the middle, the amount of the sag determines the method of repair. For a moderate sag, say $\frac{1}{4}$ to $\frac{1}{2}$ inch, a good plan is to add truss rods, one on either side. These should be stout bars, well anchored near the ends of the frame and at points where the frame has not been weakened by excessive drilling. They should be given a flattened U-shape, with two (or more) uprights down from the frame between them. The material for them should be stiff enough and strong enough to withstand bending and should be firmly fastened to the under side of the frame. The truss rods should be made in two parts with a turnbuckle to unite them, the ends being threaded right and left to receive the turnbuckle. When truss rods are put on a sagged frame, it should be turned over and loaded on the under side; then the turnbuckles should be pulled up so as to force the middle or sagged part upward a fraction of an inch, say $\frac{1}{8}$ to $\frac{1}{4}$ inch, and then the frame turned back, the other parts added, and the whole returned to use. A job of this kind which takes out the sag so that it does not recur is a job to be proud of.

Fracture. Many frames break because too much metal was drilled out at one place. Fig. 13, shows a case of this kind. The

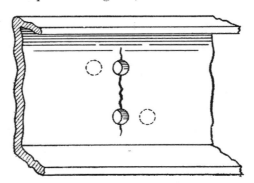

Fig. 13. Reboring Cracked Steel Channel

two holes were drilled, one above the other, for the attachment of some part, and were made too large. They were so large that at this particular point there was not enough metal left to carry the load, and the frame broke, as indicated, between the two holes and also above and below. A break of this kind can be repaired in two good ways. The first and simplest, as well as the least expensive, is to take a piece of frame 10 to 12 inches long, of sufficiently small section to fit tightly inside this one. Drive it into the inside of the main frame at the break, rivet it in place firmly throughout its length, and then drill the desired holes through both thicknesses of metal.

This is not as good as welding. A break of this kind can be taken to a good autogenous welder who will widen out and clean the crack, fill it full of new metal, fuse that into intimate contact with the surrounding metal, and do so neat and clean a piece of work that one would never know it had been broken. When a welding job is done on a break like this, and no metal added besides that needed to fill the

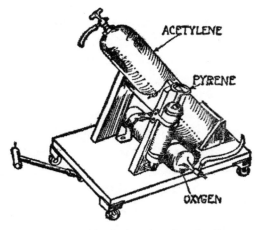

Fig. 14. Handy Oxy-Acetylene Outfit

crack, subsequent drilling should be at an angle, to avoid a repetition of the overloading condition. In this figure, the dotted lines suggest the drilling. By staggering the holes in this way, there is a greater amount of metal to resist breakage than would be the case with one hole above the other—a method which might preferably have been used in the first place.

So much welding is done now, and so many people know of its advantages, that every repair shop of any size should have a welding outfit. A frame job is essentially an inside bench job, but a large number of cases of welding could be done directly on the car outside the building, particularly in summer when the outside air and cooling breezes are desirable. So, it is well to construct a small truck on which to keep the oxygen tank, acetylene cylinder, nozzle for working, and a fire extinguisher. One form of a truck is shown in Fig. 14. This truck is a simple rectangular platform with casters, a handle, and a rack to hold the tanks. It saves many steps and is particularly convenient in summer months. This outfit is essentially a home-made affair, but the gas-welding and electric-welding manufacturing companies have designed small outfits especially for automobile repair work, which would be preferable to the one in Fig. 14, especially where the amount of repair work warrants a reasonable expenditure for a welding outfit.

Riveting Frames. *Tightening Rivets.* Rivets securing the corners of a frame or holding cross-members, gussets, and plates often work loose, particularly with the flexible type of frame previously alluded to. The location of the rivet and the accessibility of the part will determine how best to proceed with the work. The chief trouble experienced is that of placing a sufficiently solid article against the rivet while the other end is being hammered. As a rule, old axes, sledges, and hammers will serve under ordinary conditions, but these cannot always be used in a channel frame. One method is to employ an old anvil which is turned upside down and so placed in the frame that the flat end of the anvil is placed against the head of the rivet, while a rivet set is employed to set the rivet up snug. The horn of the anvil is allowed to rest on the other side of the frame. This method can be used for cutting off rivets as well as for tightening old ones. The anvil should be of sufficient length to rest on the frame as above described.

When an anvil is not available, the following method may be used with success: Take a $\frac{1}{2}$-inch bolt and cut it off so that it will just go in the frame between the rivets. Slightly countersink the head of the bolt with a cold chisel. Put on the nut and slip in between the rivets and run the nut down until it expands tight in the frame. The depression in the head of the bolt, and the nut fitting around the opposite rivet head will keep it firmly in place while riveting. It is not always practical to attempt to tighten a rivet. The better method is to remove it, drill a larger hole and use a larger size rivet. Rivets are usually made of Norway iron. Heat to a red heat before using.

Riveting Methods. There are two methods of riveting, the driving in and the backing in. The latter method is shown in Fig. 15, and the two plates to be riveted are drilled in the usual manner, as shown at *A*, with the rivets a trifle smaller than the hole, placed as shown at *B*. With hot riveting, the hole should be about $\frac{1}{16}$ inch larger than the rivet, but with cold rivets, the opening should be such that the rivets will slide in. Instead of backing up the head of the rivet, a dolly is applied to the small end, as indicated at *C*, and the driving is done on the head of the rivet by a set *D* and a hammer. The energy of the hammer is applied through the set to the rivet, which is upset or enlarged, as it is unable to move because of the mass of metal in the dolly. The metal of the rivets expands sidewise at *A* and *B*, completely filling the space. A feature of this method is that a part of the hammer blow is expended in forcing the plate *N* into contact with the plate *O*. The metal at *B* is prevented from moving sidewise by a head formed at the dolly end of the rivet, and additional blows of the hammer tend to bring the plates closer and to hold them. The backing-in method is practical in making the various styles of rivet

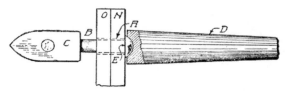

Fig. 15. Method of Riveting Frame

heads, particularly in making the thin, almost flush, head, and an advantage is that there are no reactionary stresses upon the thin head as would exist with the driven-in rivet.

As there is more demanded of the rivet replacing the old member, it is important that the work be carefully performed. This applies to the holes in the plate. All sharp corners should be removed, as they afford an opportunity for the rivet to shear off by external stress or to fly off under internal strain. A reamer, drill, or countersink can be used in removing sharp corners. The face left need not be more than $\frac{1}{64}$ or $\frac{1}{32}$ inch wide, in order to greatly strengthen the rivet at its weakest point, or where the head joins the body. By slightly chamfering the corner of the plate, the rivet is given a corresponding fillet, which not only increases its holding power but serves to draw the plates together.

Frame Bracing Methods. There are several methods whereby a frame that has been injured through collision or has sagged because of too light construction can be repaired. The front of the frame is the chief offender in this respect, and many times a leaking radiator is the result. When repairs to the radiator fail to cure the trouble, it may be assumed that the frame is at fault. A simple remedy is shown in Fig. 16 and consists in bracing the frame by means of a rod

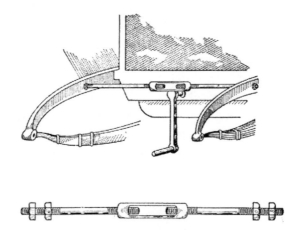

Fig. 16. Adding a Truss Rod to the Front of a Weak or
Damaged Frame to Strengthen It and Preserve
the Radiator

and turnbuckle. The rod should be about 2 inches longer than the width of the frame and threaded for about 3 inches on each end. The turnbuckle is not essential, but it simplifies the work. In

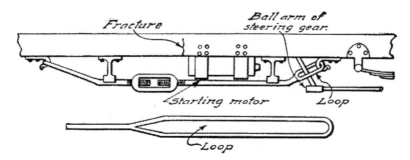

Fig. 17. Bracing Fractured Frame with Bar and Turnbuckle

installing the brace, the inside nuts are screwed on first and far enough to allow putting the rod in place. These nuts are next screwed out until they bear against the frame, and the latter is forced out until any pressure that may have existed on the radiator is eliminated. The outside nuts are then screwed up snug. The advantage of the turnbuckle is that adjustments may be made as required.

GASOLINE AUTOMOBILES

Fig. 17 shows a method of trussing a frame that was fractured by the stresses of the motor starter. Even after the fracture had been repaired, the driving gear of the starter would not mesh properly with the ring gear on the flywheel of the engine. As the movement was up and down on the frame, a truss was found necessary; while it was a simple matter to attach one end of the truss on the left-hand side of the chassis, the right-hand side was more difficult because of the proximity of the ball arm of the steering-gear lever. The problem was solved by forming a loop at one end of the truss of sufficient width and length to permit travel of the ball arm. By utilizing a turn-buckle the desired tension was obtained.

Frame Alignment. It is often found that the wheels of a car do not run in track with each other. This trouble can be readily

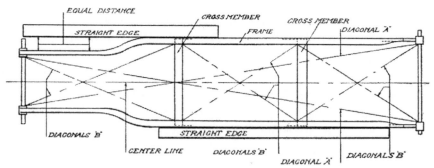

Fig. 18. Frame Alignment

seen if a person will stand behind or drive behind a car that has this trouble. It will be noticed that the rear wheels do not run in the same track as the front wheels, while they seem to carry the rear part of the car to the side of the road causing the driver to turn the front wheels to keep the car moving ahead. This condition can be caused by two things: The axle may have shifted on the springs or the frame may be out of alignment. To check the position of the axle, measure the distance between some point on the frame, such as a stationary spring bolt and the center of the axle. If the axle is in the correct position, the distance measured will be the same on both sides of the car within one-eighth of an inch. The distance should be the same between the front and rear axles on both sides of the car when measured between the centers of the axles. If the measurements do not check the same on both sides of the car within the allowance measured, the frame should be carefully inspected for bends. Sometimes, in an accident, one side of the frame will suffer all the damage and will be driven back out-of-alignment with the other side. The first thing to do when putting a frame in such condition into alignment is to divide the frame down the center into two parts with a line, Fig. 18. The long diagonals, marked *A*, will be the same length if the frame is in correct alignment. The short diagonals, marked *B*, as measured between the cross members of the frame, will also be of the same length. It will be noticed that in each case the diagonals cross on the center line of the frame. The side members of the frame may be bent. This condition can be checked with the aid of a straight edge, Fig. 18.

Worn Spring Hangers. Through neglect, the spring hangers and spring horns may become badly worn where the spring bolt passes through them. Cut them free from the frame, fill the holes by welding and redrill the bolt holes. The parts should then be riveted back on the frame.

Worn Rivets and Rivet Holes. When rivets and rivet holes are worn, and it is not possible to fit new rivets, repair can be made as follows: Cut out the old rivets, drill the holes to a bolt size that is a little larger than the rivet so that a bolt will be a light driving fit in the hole. The nuts should be put on the bolts with a lock washer under them and pulled up tight and then the bolt ends riveted over. Special care should be taken to see that the bolts are a good fit in the hole.

SPECIAL TYPES OF DRIVE

Front=Wheel Drive. In the conventional type of pleasure motor car, the energy of the engine is applied to the rear wheels which propel the car, the drive being a pushing one. A pleasure car, or rather a racing machine, with a front-wheel drive—which is a pull, and held by some to be more economical—was brought out several years ago but not marketed. During the latter part of 1916, a company was formed to market an eight-cylinder pleasure vehicle, utilizing a front-wheel drive and steer and a friction drive with an automatic pressure control.

Difficulties of Transmission. The Homer Laughlin car was one of the first cars that made use of an original type of universal joint to transmit uniform angular velocity. Its design was brought about by the fact that the rate of transmission of angular velocity through a universal joint is not even when the shafts are at an angle. This is the fundamental difficulty every designer of a front drive has to overcome or suffer the twisting of the axle.

The front wheels and the flywheel must rotate at practically a uniform speed, at least through each revolution. The irregular rate of transmission through the universal joint must be taken up some-

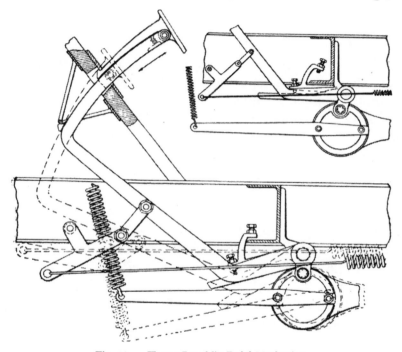

Fig. 19.   Homer Laughlin Pedal Mechanism

where. The normal action of a universal joint at certain angles is to make four jerks in a revolution, as it has four fast points and four slow points. The Laughlin joint gives uniformity of rotation with 75 per cent on each side of normal, the difference being taken up by the flexibility of the transmission parts.

Friction-Disc Transmission. The transmission is of the friction-disc type, but the disadvantage of this form of drive—the fact that the control is reversed—is eliminated. The usual clutch control is provided, but the pressure is automatic. This pressure is obtained by an eccentric connection by means of which designers obtain irreversible application of spring pressure. The transmission locks at the correct pressure through the friction of the eccentric. The spring controlling the friction for driving provides the proper pressure for running, but it is not sufficient for starting or climbing long hills in the low gear. The pedal shaft operates a dog that presses down on the eccentric sheave extension. To de-clutch, the operator presses the pedal down, releasing the clutch. The pedal has two points at which it latches, providing extra pressure, and an extra spring is brought into service for the high and low speed. This spring operates through a toggle linkage. As the pedal rises, the applied power increases. When the car attains momentum, the driver depresses the pedal until it latches. The running pressure is sufficient to hold the engine in all gears except the low and reverse.

Control. Complete control is obtained through one gearshaft, the lever working forward for progressive, and back for reverse. Automatic latching is obtained in every gear, the latch working in sockets sunk in the jackshaft. Chain drive is employed between the transmission and front axle. The brakes are located on the rear axle. Fig. 19 shows the method of obtaining a conventional pedal control of the transmission through the irreversible application of spring pressure—one spring for ordinary service, the other for low gear work—controlled by the eccentric on the jackshaft of the driving mechanism.

Four=Wheel Driving, Steering, and Braking. The four-wheel drive—a construction in which all four wheels of the vehicle drive, and frequently steer and brake—is confined to commercial vehicles. A brief consideration of the actions which may have to take place at the same time in such an axle will give a very good idea of the problem which must be worked out. The wheels must be free to turn about the axle as an axis, being driven from their hollow centers; the wheels must also be free to turn about the pivot point as an axis swinging in a horizontal direction and must be driven steadily all the time. All the turning, swinging, and driving action must be outside of and beyond the spring supports of the chassis, since the body cannot turn; but the axles must at the same time support the springs. Further, if all four wheels are to carry brakes, they must be applicable at any and all times and at any and all angles of inclination of the wheels, either in a vertical or horizontal direction, and they must be so equalized as to apply equally to all wheels, no matter how the force is applied to the system, and no matter in what position the wheels may be.

The advantage of the four-wheel drive and with it the four-wheel steer and brake is granted by eminent engineers, as is also its

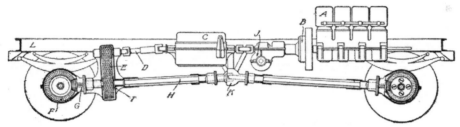

Fig. 20.   Side View of a Four-Wheel Drive, Steer, and Brake Motor Truck

necessity for heavy commercial trucks, but its use has not been extensive for the simple reason that it is a complicated arrangement at best.   In many cases, the design has been so complicated and unmechanical as to cause failure, and the reports of these troubles have given the four-wheel driving, steering, and braking device a sort of visionary air, so that any one talking of it is supposed to be a dreamer. Such is not necessarily the case, for many different practical four-wheel combination driving, steering, and braking devices have been brought out, built, tested, and proved efficient.

A number of four-wheel designs for commercial cars are being marketed, and have proved the contention of their makers that they are economical in operation and maintenance.

Four=Wheel Steering Arrangement.   With the design shown at Fig. 20, steering knuckles are eliminated, the wheels being con-

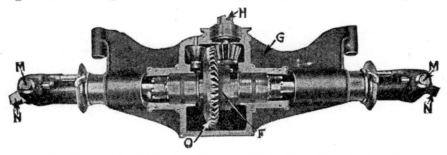

Fig. 21.   Details of Axle of the Four-Wheel Drive Truck Shown in Fig. 348.

nected to the axle ends through the medium of vertical trunnions. These trunnions bear on the wheel ball-bearing ring, which is ample in diameter and turns freely because of its size and the use of ball bearings. Within this ring, the axle terminates in what is practically a universal joint, driving through to the outside of the wheels. The wheels are thus free to run about a point in the axle ends, at the same time taking their power through the inside rotating shaft. Fig. 21 illustrates one of these axles with the parts lettered. Here H is the point of attachment of the driving propeller shaft, G the cast-steel one-piece case, F the differential gear within the large driven bevel gear O, MM the vertical trunnions upon which the wheels rotate, and NN the universal joints which drive the wheels.

How the steering is obtained is shown in Fig. 22. At the front of the chassis is the steering wheel P; turning it partially rotates the longitudinal shaft Q, which extends the length of the chassis. This shaft carries levers RR near its two ends, which are connected to

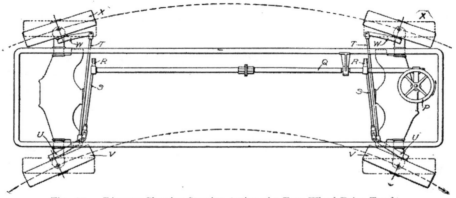

Fig. 22. Diagram Showing Steering Action of a Four-Wheel Drive Truck

the steering rods SS. These rods connect to the steering levers UU, which are fixed to the wheels themselves instead of to the steering knuckles as in the ordinary case, for this car has no steering knuckles. In addition to the steering rods attached to the longer of the two steering levers, there is a cross-connecting rod TT at each end, which con·· nects the two steering levers. Thus, when the levers RR move the rods SS, and through these the levers UU, which in turn move the wheels VV, the rods TT also come into play and move the levers WW and the wheels XX. Therefore, the movement of the steering wheel in any given direction, as to the right, turns all four wheels, the front two to the right, and the rear two to the left so that they form arcs of the circle in which the front ones are turning. The truck thus makes the desired turn to the right in one-half the distance or time of the ordinary truck. Four-wheel steering then has the advantage over two-wheel, or ordinary, steering, of requiring only one-half the space and one-half the time to accomplish a given turn. The vehicle described would turn completely around in a circle of 40 feet, the outermost circle shown in Fig. 22 being 56 feet in diameter.

Chain Four-Wheel Drive. Fig. 23 clearly illustrates at bottom view of the Hoadley four-wheel drive, four-wheel steer, and four-wheel brake truck. The power of the engine is transmitted through shafts, gears, and universal joints to the differentials; there is a third differential in the gear box at the center of the frame. Final drive is by chain; both ends of the truck are exactly alike in so far as the four-wheel drive is concerned, and the fifth wheels run in ball bearings. Steering is accomplished by means of worm gearing, the shaft being clearly shown, and both sets of wheels are steered simultaneously.

Jeffery Quad. An example of the successful development of the four-wheel drive is the Jeffery Quad, Fig. 24, which has given an excellent account of itself in governmental work. In this type it will be noted that the inclined driving shafts, shown in Fig. 20,

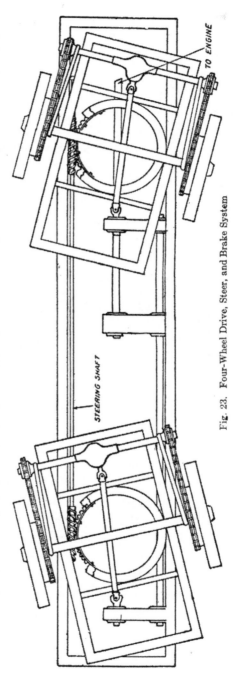

Fig. 23. Four-Wheel Drive, Steer, and Brake System

have been carried up to the gear box with a universal joint on either side. This construction has resulted in a much more inclined shaft in each case, but it has also eliminated the tail shaft D, the use of a silent chain E with its housing, the central universal joint, and the spherical bearing K, and, in addition, it has simplified both shafts.

In the four-wheel drive vehicle the engine was placed on the center line of the car; on the Jeffery it is set off to one side, while the two driving shafts to the front and rear axle, which form a continuation of each other, are set off to the other side. This result is produced by making the transmission very wide with three side-by-side shafts, as shown in Fig. 25. The engine drives the splined shaft H, on which are gears that transmit the rotation to the intermediate shaft C, which through the final gears E and F, drives the final shaft, which is in two

Fig. 24. Plan View of the Jeffery Quad, Showing Disposition of Units
Courtesy of Thos. B. Jeffery Company, Kenosha, Wisconsin

parts, B driving one pair of wheels, G driving the other pair. Note that the differential has been incorporated in this type of drive, so that it is possible to have a different drive for the front wheels from that for the rear wheels.

The rest of the construction is too simple to require a detailed description beyond the simple statement that the gear box gives four forward speeds and one reverse. When the two ordinary shifters are in the neutral position shown, reverse is produced by shifting the double reverse gear on shaft D along until its left-hand member meshes with the second-speed gear on shaft A and its right-hand member with the low-speed gear on shaft C.

Universal joints fit on the two tapers B and A with shafts inclined to the two axles. On top of the stationary axle of the I-beam section is fixed a small box which contains the bevel gears and an additional differential with suitable bearings, the whole being enclosed. These can be seen in Fig. 24, that on the rear axle being

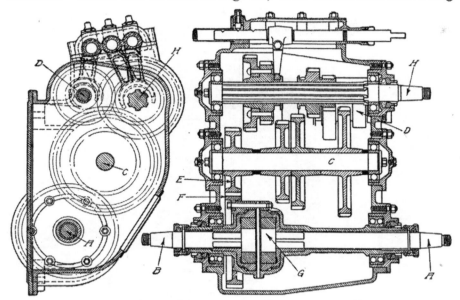

Fig. 25. Plan View of the Transmission of the Jeffery Quad, Showing the Shafts for Both Axles

plainly shown, while the one in front is partly obscured. This member is shown in detail in Fig. 26, which gives the longitudinal section along the driving shaft at the left, in which the axle H is noted, the bevel gear I, and the bearings for radial and thrust loads

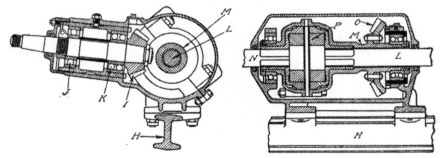

Fig. 26.   Sections Showing Bevel Drives at the Axles on Jeffery Quad

at J and K, respectively. The driven shaft is seen at L, with the sleeve M around it, the sleeve being used to drive to the differential case, since the larger, or driven, bevel C is not sufficiently large to house the differential P.

Fig. 27 is a diagram showing the details of the axle end and wheel construction. In this, H is the I-beam section of the axle bed shown in Fig. 24, and N one of the shafts, which carries at its end the universal joint Q, with the end of the shaft extending beyond the joint R. The latter carries the spur gear S, which meshes with

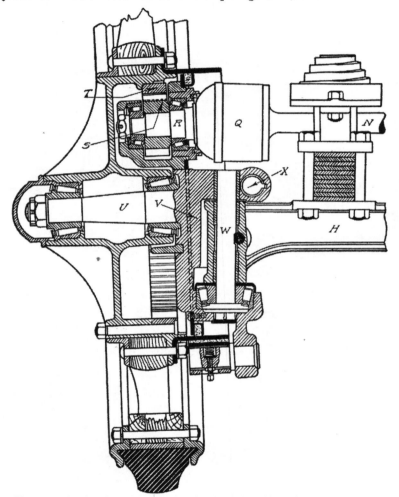

Fig. 27. Section through a Wheel and Axle End of the Jeffery Quad, Showing Method of Driving and Steering

the internal gear T fixed to the wheel and drives the vehicle in this manner. It will be remembered that this is not necessarily a front wheel, but any one of the four.

The wheel turns on the spindle U, which is part of the steering knuckle V; this knuckle turns upon the pivot W. The lever which turns the wheel is attached at X, the pair (either both front or both rear wheels) being connected by means of a cross-rod; at one end of this rod there is a connection to a rod which runs the entire length of the chassis. This rod is operated by means of the steering gear, and imparts the same motion to the front wheels as to the rear, except that the two are in opposite directions, that is, front wheels turn to the left and rear wheels to the right, so that they will follow around in a correct circle.

Advantages of Four-Wheel Drive. It is claimed for the four-wheel drive that its four-wheel steering reduces the mileage traveled to the minimum in that the car can run closely to corners and travels less in crowded traffic, in turning around, and in approaching and leaving loading platforms. The push of the rear wheels and pull of the front wheels enables it to surmount obstacles instead of bumping over them, and its greater traction permits it to travel soft roads not easily negotiated by the rear-drive type of trucks and cars. The four-wheel drive type will turn in a 48-foot circle, and, with its locking differential, obtains traction on slippery roads.

Electric Drive. When the final drive is electric, or when the source of power is an electric motor, the matter of four-wheel driving is much simplified, the wheel carrying the electric motor attached directly to it and turning with it about the knuckle pin. Both wheel and motor are turned by means of a worm and gear above, the wheel being attached to the upper end of the steering-knuckle pin prolonged. Turning this turns the wheel and motor.

This steering wheel is turned by the worm, which is on one end of a cross-shaft. This shaft is carried in bearings above the stationary bed of the axle and has near the center a bevel gear that meshes with another bevel, which is, in turn, attached to the lower end of the steering post. Turning the steering wheel turns the post and the bevel gear, which turns the bevel pinion and with it the worm shaft. The shaft turns the worm and the worm wheel which actuates the road wheels. The driver thus has a triple reduction between himself and the wheels, giving him this much advantage in steering: there is the leverage of the wheel of large diameter, the ratio of the sizes of the two bevels, and the ratio of reduction of the worm gearing, which, in addition, is irreversible. The steering gear is thus eliminated and four simple gears substituted for it.

Couple-Gear Type. In the Couple-Gear wheel, which is an American product, the motor is placed inside of the wheel—a type especially designed and constructed for this purpose. With the motor in this position, the wires enter through the hollow hub, altering its construction very materially. As compared with the electric motor on each wheel, previously described, this form has the advantage of greater simplicity, fewer parts, superior appearance, and protection against the elements, while the enclosed position of the motor, which is the most delicate part of the machine, protects it against road obstructions and accidents. This arrangement also simplifies the steering problem, since the car is steered just the same as any other truck, much of the complication incident to an electric motor on each wheel being eliminated.

Fig. 28.   End View of Couple-Gear Electrically Driven Wheel with Tire Removed

Fig. 28 is a view of the wheel with the tire removed and the whole disconnected from the axle ends. Aside from this, it is complete and ready for use. Note how the axis of the motor is set at a very slight angle, just sufficient to allow a pair of very small driving gears at the two ends of the armature shaft to drive on opposite sides of the wheel. The wheel is assembled with a pair of driven gears on either side, these being separated a comparatively small distance, about $2\frac{1}{2}$ to 3 inches. As stated, the armature shaft has a small bevel pinion on each end, each of these meshing with the driven gears, but on opposite sides. It is this arrangement which gives the device its name of Couple-Gear. In this figure the brake band has been removed, but the brake drum will be seen just inside the wheel at *A*. Beyond this is noted the spindle *B*, which is made hollow for the wires from the battery and turns in a bearing on the axle.

In the second illustration, Fig. 29, an axle, either front or rear, with the wheels removed, is presented. In this cut the left wheel is entirely removed, but the one on the right shows the axle spindle *B*, the method of fixing it in the axle support at *C;* the armature housing *D* is normally within the wheel and not visible. One feature peculiar to this arrangement is the steering, which is effected by means of a vertical post with a small spur gear at its lower end *E*. This meshes with a curved rack *F*, which is machined on the outside of a pivoted member *G*, to which a pair of arms are attached. One of these arms *H* has a rod *I*, which runs to and operates the right-hand spindle *B*, while the other *J* has a similar rod *K*, which operates the left-hand wheel. When all four wheels are to be driven in this manner, the post is vertical, but the connection with the rack *F*

Fig. 29. The Couple-Gear Axle and Parts, Showing Method of Operation
Courtesy of Couple Gear Freight Wheel Company, Grand Rapids, Michigan

becomes horizontal, with a continuation to the rear axle which operates the various arms, levers, and rods there in the same manner.

This particular system is used for heavy commercial work only, and in this it has been particularly successful as a tractor, a front axle and a pair of wheels being substituted for those of a heavy trucking wagon. Then, with a sling under the body or beneath the driver's seat for the batteries, and with proper wiring, control levers, and steering wheel, the truck becomes electrically driven.

275 H. P. HEAVY-DUTY PACKARD MARINE MOTOR
Courtesy of Packard Motor Car Company

SPRINGS AND SHOCK ABSORBERS

SPRINGS

Basis of Classification. Springs may be divided into four general classes as follows: semi-elliptic, the full-elliptic, the three-quarter elliptic, and the cantilever. The full elliptic spring is made up of two sets of flat plates, slightly bowed away from each other at the center and attached together at the ends. When these are used, the centers of the springs are attached top and bottom, respectively, to the frame and axle. With half of the top of the spring cut away, and the cut, or thick, end attached to the frame, this spring becomes a three-quarter elliptic. When the whole top of the spring is cut away, so that the spring is but a series of flat plates, bowed to a long radius, this becomes a semi-elliptic spring. By turning the semi-

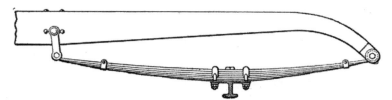

Fig. 1.   Typical Semi-Elliptic Front Spring

elliptic spring over, it becomes a cantilever when its center and one end are attached and the load applied to the other end. The coil form requires no explanation and is not now used on cars. In addition, these forms are modified by scroll ends and various attachments.

Semi=Elliptic. Fig. 1 shows a front spring of the semi-elliptic type, the form which is used now for almost every front spring. This is a working spring of the usual type, fixed at the front end, shackled at the rear end, attached to the axle in two places, and with two rebound clips in addition. The latter are put on the springs to prevent them from rebounding too far, in the case of a very deep drop. In some cases, as high as four, six, or eight of these clips may be used. Many other springs are made with ears, these being clipped over the next lower spring plate, the final result being the same as the use of many clips, but with improved appearance.

Full=Elliptic. Full-elliptic springs are the oldest form known. Fig. 2 shows the construction of this type, the upper and lower parts

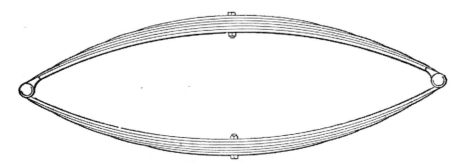

Fig. 2. Typical Full-Elliptic Front or Rear Spring

being pivotally connected at the ends. A slight modification of this form, known as the scroll-end full-elliptic type, is in more extensive use than the full-elliptic plain type. As Fig. 3 shows, the ends of the upper leaves are bent over. Each carries an eye, which is connected to

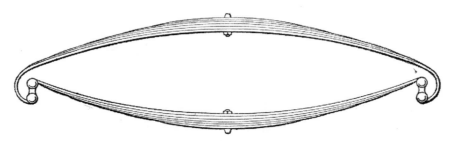

Fig. 3. Full-Elliptic Spring with Scroll Ends

the eye in the end of the upper leaves of the lower half of the spring by means of a shackle. This construction makes a very soft-riding spring.

Three=Quarter Elliptic. Very much like Fig. 3 is the form known as the three-quarter elliptic spring, the one having scroll ends being shown in Fig. 4. This form of spring is fastened at three points. The lower part of the spring is shackled at the front end, fixed to the axle at the center, and shackled to the upper part of the spring at the rear. The upper part of the spring is fixed to the frame at the upper front end and shackled to the lower part at the rear. Fig. 5 shows another example of the three-quarter elliptic spring, which may differ in practice, as some three-quarter springs are not scroll ended.

This form of spring is growing in favor daily, a greater number being used this year than last, while designs for next year show a still

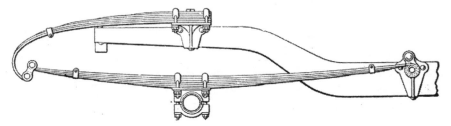

Fig. 4.   Three-Quarter Elliptic Rear Spring with Scroll Ends

greater increase.   One reason for this increase is the great increase in the number of dropped frames, that is, frames unswept at the rear.   To

Fig. 5.   Three-Quarter Scroll Elliptic Springs on Winton Car

this form of frame, the three-quarter elliptic spring is very well adapted and makes a very natural, very good, and very easy-riding combination.

Platform.   The platform type of spring is used a great deal on large cars, as well as on very heavy trucks, on account of its ability to carry heavy loads well, and also on account of its flexibility.

As may be seen in Fig. 6, it consists of three semi-elliptic springs shackled together at the corners. The rear cross-spring is usually made shorter than the two side springs, while the latter are set off center, making the front of the spring, that is, the part forward of the point of attachment to the axle longer than the part to the rear. There are two reasons for this: First, the front end acts somewhat as a radius rod, the rear end of the frame rising in an arc of a circle

Fig. 6. Platform Springs, Showing How Side- and Cross-Springs Are Shackled Together

whose radius is the front half of the spring; second, this plan distributes the spring action equally in front of and back of the axle. Since the rear cross-spring is fastened to the frame in the center,

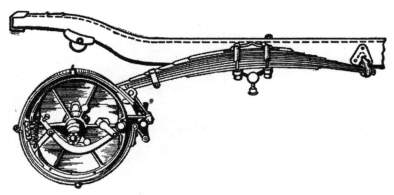

Fig. 7. Cantilever Rear Spring Used on King Cars
Courtesy of King Motor Car Company, Detroit, Michigan

each half of it is considered as a part of the side spring to which it is shackled. Thus, the total length of the side spring in front of the axle is the measured length of the side spring, while the total length of the side spring back of the axle is considered as the side length plus half of the cross-spring length. The center point, or point of axle attachment, is not moved so far forward as to make these two lengths equal, but in a proportion which may be derived thus: Assume a side spring 42 inches long and a cross-spring 35 inches long; then the spring would be set out of center some 4½ inches, making the front length about 25½ inches, while the rear length would be 16½ inches

Fig. 8.   Front End of Cantilever Spring on Siddeley-Deasy (English)Car

plus half of the rear spring, or 17½ inches, making a total of 34 inches. This would give a ratio of 25½ to 34, or 1 to 1.333.   If the side members were 50 inches, the ratio would be about 1 to 1.25, and for side members shorter than 42, the ratio would be about 1 to 1.5.

Cantilever. The cantilever is, in appearance, a semi-elliptic spring turned over.   It gets its name, however, from the method of suspension, which is quite different from that of any form of semi-elliptic spring.   Moreover, as a part of this suspension, at least one end of the cantilever and sometimes two are finished up flat and square to slide back and forth in a groove provided for that purpose, a bolt through a central hole preventing the spring from coming out of its guide.   One form, shown in Fig. 7, has a fixed attachment to the rear axle, a pivoted attachment to the frame at its center (or slightly beyond the center), and a sliding attachment to the frame at its forward end to take care of the increase in length and of the forward movement necessary when the rear wheels rise.

Another form of cantilever is that shown in Fig. 8.   This is the rear spring on the Siddeley-Deasy (English) car, and like that of the King, is pivotally mounted on the frame just forward of its center. Unlike the King, however, the forward end of the spring has a shackle which permits it to swing when the rear axle rises or falls.   This shackle is a very interesting feature of this installation, having an adjustment which is most unusual for a shackle, Fig. 9. Note how the outsides of the shackle have a series of grooves, into which the head of the shackle bolt on one side and the washer on the other, fit. By setting these in the desired grooves and tightening the nut, the position is fixed. If this does not give the proper throw, it is a simple matter to remove the nut and make a new adjustment.

In France, a form of double cantilever has been tried out with success; this form consists of a pair of cantilevers, one above the

Fig. 9. Detail of the Adjustable Shackle on Siddeley
Cantilever Spring

other, separated at the center by a carefully sized spacing block, which is pivotally attached to the frame. The rear ends are attached above and below the axle, while the front ends are attached to two fixed points. Although the ends are made much thinner and more flexible than those just shown, it should be noted that both of them are fixed. The rise and fall of the wheels must be taken up by the springs themselves, the pivot in the center simply distributing the distortion over both the front and rear halves.

Advantages of Cantilever. The advantages of the cantilever spring are the smaller unsprung weight and the reduced manufacturing cost for a given amount of flexibility. Other advantages are the absence of sharp rebounds and a greater deflection for a given load and length of spring; it also obviates the cut in the body required with the three-quarter elliptic spring. When the cantilever takes the driving strain, the main leaf is usually stiffened and, being stronger sidewise, it eliminates a good deal of the side sway. With torque rods, the main lead may be made lighter, as the starting, the braking, and the torque act through the torque rods. Since there is more metal in the line to the thrust, they are especially suitable for taking the thrust, and not quite as efficient in taking the torque.

Hotchkiss Drive. The adoption in 1915 of the Hotchkiss drive, Figs. 5 and 10, in which the rear axle is connected with the frame through the chassis springs only, making the springs perform the

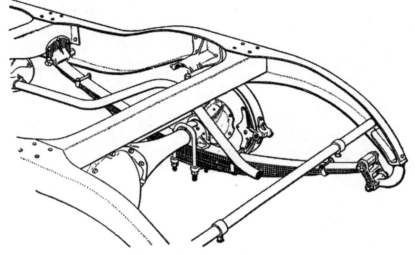

Fig. 10. Rear-Spring Underslung

functions of torque and thrust, is a radical departure from previous forms. The objection that it subjected the springs to unnecessary strains has not been sustained in practice, which has shown that a slight yielding of the rear axle when starting and braking, by a certain flexure in the springs, has reduced the stresses upon the transmission members.

In the Hotchkiss drive, the springs are rigidly attached to the rear axle, while the front end of the spring is secured to the frame with a proportionately large bolt through which the drive is transmitted. Users of the drive claim that it is quieter, that the car holds the road better, that it is more flexible, and that it avoids the road shocks which are transmitted through stiff torque members from the axle to the frame. Makers who drive through the springs and employ other torque members claim that they are not sacrificing flexibility in driving while eliminating a certain side sway and other strains prevalent when the springs perform the functions of the torque. In the Hotchkiss drive, two universal joints in the drive shaft are used.

Unconventional Types. *Marmon.* A departure from conventional practice is the spring used on the Marmon car and shown in Fig. 11. It is a double-transverse construction, consisting of semi-

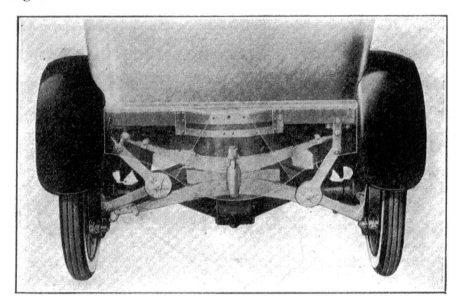

Fig. 11. Unique Rear Spring of Marmon Cars

elliptic springs bolted together at the center, with a curved block, or hard-maple cam, between them. This cam varies their stiffness, the spring automatically becoming stiffer as the load increases. Under normal load, the stiffness is about 170 pounds per inch, but as the springs are compressed the stiffness will reach 400 pounds. They are shackled at one side and fixed at the other, obtaining a perfectly parallel motion to the frame. There is said to be no roll as is sometimes found with transverse springs.

Knox Tractor. An unusual method of suspension is that employed on the Knox tractor, a combination of a cantilever and semi-elliptic spring at the rear end of the frame. The design shown in Fig. 12 includes heavy semi-elliptic springs, which are attached to the rear axle by long clips and carry the fifth wheel of the trailer. There is no connection between the springs and the tractor frame, so they carry the weight of the trailer and load only. The tractor frame

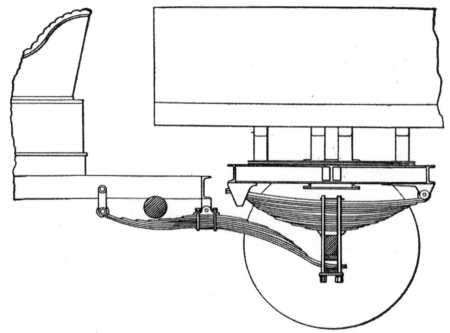

Fig. 12. Combination Cantilever and Semi-Elliptic Spring on Tractor

is mounted on a cantilever spring having a pivot near its center and a shackle at the front end. The rear end bears on a seat clipped to the

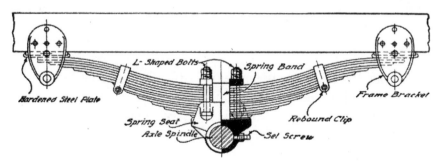

Fig. 13. Rear Spring of Six-Ton Truck

rear axle. This obtains a flexible mounting for the tractor and also permits the carrying of very heavy loads on the trailer.

Semi-Elliptic Truck Spring. The semi-elliptic spring is a favorite with makers of commercial vehicles. It is simple, and if the length, width, and other dimensions are proportioned correctly, it is a most satisfactory method for both front and rear suspension. Fig. 13 shows a rear spring for a 6-ton truck, the method of shackling, and how it is mounted on the axle by means of a spring seat.

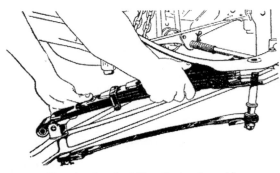

Fig. 14. Overland Four-Spring Assembly

Many makers are now using their own special form of springs. Fig. 14 shows the spring arrangement used on the Overland Four, a type that may be described as a three-point suspension. It was made in two parts, the lower part consisting of a regular semi-elliptic flat spring, while the upper part was a semi-elliptic flat spring with scroll ends. The central part of the spring was treated as one, being attached to the axle in the usual manner; the ends, however, had a peculiar appearance, because the upper and lower halves of the spring were of different shape. The scroll end of the upper part was supposed in itself to absorb many of the small road shocks. The spring was loosely attached to the frame at each end by means of a double shackle, made necessary by the double action of the spring; the tendency to flatten out increased its length, thus calling for a forward motion of the front and a backward motion of the rear ends, while the different lengthening action, owing to the difference in the lengths of the two parts of the spring itself, resulted in a turning about a different point.

In the latest form of spring construction on the Winton car, Fig. 5, it will be seen that the three-quarter elliptic form has been adopted, with a kick-up at the rear end of the frame. If the two types are compared somewhat closely, it will be seen that the only change in the frame part is the kick-up. The new springs show the scroll ends to which Winton has always been partial.

Ford. The form of the Ford spring has always been distinctly different. Fig. 15 shows the front and Fig. 16 the rear spring used on Ford cars, the distinction in the front spring being principally in the use of a single ordinary inverted front spring set across the frame on top of the axle, where most makers use a pair of side springs set parallel to the frame. This form is simple and cheap to make and assemble, the cost of the spring itself, and the work of putting it on being just about half that of the spring attachment of the

Fig. 15. Special Vanadium Front Springs for Ford Cars
Courtesy of Ford Motor Company, Detroit, Michigan

ordinary two-spring type. On the other hand, excellent riding qualities are claimed for it. A second distinction is that the spring is an inversion of the usual semi-elliptic type, the set of the spring being downward instead of upward. A third claim to distinction is in the use of vanadium steel, which, it is claimed, has a higher tensile and comprehensive strength than any other steel, and it is practically unbreakable in torsion. This steel is also being used in many other

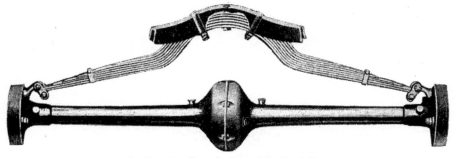

Fig. 16. Rear Springs of the Ford Car

parts, such as crankshafts, camshafts, fender irons, frames, driveshafts, etc., resulting in a very light-weight car, since the greater strength of the material allows the use of smaller sections for equivalent strength.

The Ford rear spring has all the claims to distinction of the front spring, and, in addition, a hump at the center. Fig. 16 shows this hump clearly, the rear-frame cross-member being only partly shown. It will be noted that both ends of both springs are shackled, the construction necessitating it. These springs represent quite a radical departure, the success of which has been proved in actual practice.

213

Locomobile. Fig. 17 shows the three-quarter scroll elliptic rear spring used on the Locomobile, also the method of shackling both ends

Fig. 17. Three-Quarter Scroll Elliptic Spring Used on Locomobile Cars

of the spring, and the use of a considerable extension beyond the spring clip of the two upper leaves. Fig. 18 illustrates the Locomobile front springs, the upper spring being used on the 1916 model, and the lower one on the 1917 model. As may be noted, the later type is 2 inches longer and also flatter, and the distance between the spring

Fig. 18. Two Sets of Front-Axle Springs on Locomobile Cars

bolt and eye of the shackle is less in proportion to the 1916 design. It was found that the jerky action and fore-and-aft pitching of the axle were eliminated by this construction, greatly improving the riding qualities of the vehicle.

Electric Car Springs. The spring suspension of electric pleasure cars is similar to that of the gasoline vehicle, semi-elliptic suspension in front, and full-elliptic scroll-end suspension at the rear. The method of shackling is similar.

Varying Methods of Attaching Springs. Springs are attached in many ways. For example, the one shown in Fig. 5 might be shackled at the front end, fixed to the axle, and fixed to the center of the frame at the rear, the side and cross-springs being shackled together. Again, the front end might be fixed to the frame, Fig. 19, all other connections being unchanged. Or, with either method of fixing the front end, the spring might be swiveled on the axle, so as to be free to give sidewise without changing the other properties of the spring. Or, with either method of fixing the front end of the spring, and with or without the axle swivel, the cross-spring might be pivoted at the central point so as to be free to turn in any direction

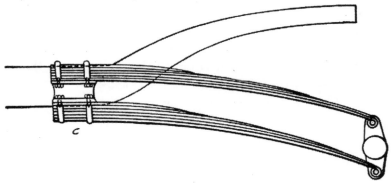

Fig. 19. Special Type of Double Quarter-Elliptic Rear Spring

about this central point. This latter method prevents binding and unequal spring action when one side of the frame is unduly raised or depressed, the solid method of fixing the rear end resulting in a double action on the part of one spring, owing partly to the tilting of the body and partly to spring action itself. With the pivot joint, the spring first swings about this point until a position of equilibrium is established, when the suppleness of the spring comes into action, the result being a deflection of half what it would be in the other case.

This form of spring also is used with the spiral spring, the latter taking the place of the shackle between the side and rear members. In this position it serves two purposes: (1) as a connector, taking the place of and doing the work of a shackle, thus acting as a universal and swinging joint between the two springs; (2) as a shock absorber, taking up road shocks within its length, that is, in the coils, without transferring any of them to the body proper or, in case of heavier shocks, sharing with the side and rear springs. This, of course, is the true function of the springs —to allow the road wheels to pass over the inequalities, rising and falling as may be necessary, while the body travels along in a straight line, level and parallel with the general course of the road.

Underslinging. Any of the spring forms shown and described may be underslung, that is, attached to the axle from below. Fig. 10. This is a quite common practice for semi-elliptic springs when used in the rear, but it is very uncommon for front springs. Similarly, full elliptics, whether having scroll ends or not, are frequently underslung. The three-quarter elliptic form when used in the rear is usually underslung; the platform spring is not underslung as often as the others. The cantilever and quarter-elliptic springs have been mentioned in connection with the underneath attachment. It should be pointed out that the position beneath the axle lowers the center of gravity by an amount equal to the thickness of the spring plus the diameter of the axle plus twice the thickness of the attaching means, and this, too, without interfering with the quality or quantity of the spring action. In the case of the cantilever, the effect of underslinging is to reduce the straightness of the spring, that is, the form when attached above the axle is almost straight, while the form when fastened below the axle is very much curved and has considerable "opening".

Shackles and Spring Horns. Considerable improvement has taken place in the method of shackling springs, and provision is now made with most types of springs for the adjustment of the shackles and hangers as well as for renewing bushings. Reference has been made to the tendency of design in rear-spring suspension and to the underslung types. Shackles are used for connecting the ends of the springs to the extensions.

There are several improvements in the method of spring attachment which tends toward longer life and service in the spring bolt and bushings with the reduction of noise and increased riding comfort. In some cases, the spring shackle bolts are entirely enclosed and thoroughly lubricated, while in others the spring bolt is entirely eliminated. In the latter, rubber blocks, between which the main spring leaf slides, take the shock and do away with all noise and rattle that may be caused by worn spring bolts as well as giving the better riding qualities before mentioned.

Departing from the conventional shackle was the safety double shackle used on the Rainer 1000-pound capacity delivery car, shown in Fig. 20. In addition to the main eye on the main leaf of the rear spring, the second leaf is extended and formed into an elongated eye, allowance being made for deflection under load. The eye of the leaf is attached to the frame by the usual rigid spring bolt. Additional means of support are furnished by clamps on either side of the spring, one by a pin through the elongated eye, and the other by a pin through the lower end of the clamp which takes in the third and fourth leaves. It is pointed out that in case the main leaf breaks, the eye of the second becomes the driving eye, and should this break, the spring will wedge between the under pin and the

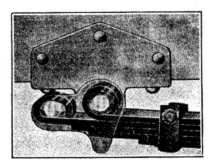

Fig. 20. Double Shackle Used on Rainer
Delivery Car

upper part of the clamp, thus obtaining rigidity which is essential with the Hotchkiss method of drive.

Although the general pratice is to shackle the semi-elliptic front spring at its rear, a departure which places the shackle at the spring horn or in front is noted in the Manly truck.

Adjusting Spring Hangers. Automatic adjustments for spring bolts are becoming very popular. The Chandler Company use a spring of the coil type which is placed between the spring bolt nut and the shackle plate. As the shackle plate wears, the play is taken up by the spring pulling the bolt and bringing the shackle plate close against the spring and hanger.

The type of front-spring hanger, shown in Fig. 21, is adjustable This adjustability is accomplished by relieving the body of the grease cup and screwing in the slotted bolt which eliminates side play,

The grease cup body acts as a lock nut. The rear hanger of the front spring, Fig. 22, is adjusted by loosening the inside lock nut and the body of the grease cup. After removing the cap of the grease cup, the hanger bolt is turned out, or to the left, with a

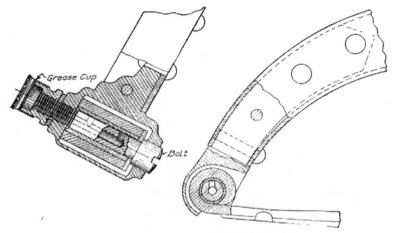

Fig. 21. Section of Adjustable Front-Spring Hanger

screwdriver, decreasing the distance between the links. The grease-cup body and lock nut are then set up tight. Provision is made with some types of rear springs for eliminating play when the rear ends are mounted on seats.

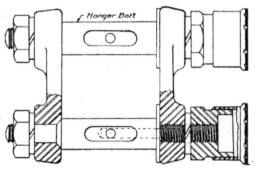

Fig. 22. Section of Rear-Spring Hanger

Spring Lubrication. All springs now are fairly well lubricated. All shackles are provided with grease cups, and other points of attachment to the frame are provided with oil holes. Where the springs are pivoted either on frame or axle, a big grease cup is usually furnished.

In addition, it is now realized that the maker can prevent much of the noise formerly coming from dry and perhaps rusted steel spring plates working over each other. There are several ways in which oiling is accomplished. The springs are made with an internal lip, or groove, which is filled with lubricant when they are assembled; or between each pair of spring leaves is placed an insert having a series of oil pockets throughout its length, each filled with lubricant normally held in by means of membrane cover; the movement of the spring plates and the heat generated thereby starts the lubricant flowing to all parts. An even later method is the attachment of external cups, provided with a wick which goes around the spring leaves and is pressed against their sides. The wick is kept wet with lubricant from the cups, and the motion of the spring leaves, together with the capillary action in the wick, draws the oil in between the leaves.

Spring Construction and Materials. A study of the illustrations used will show that practically all modern springs are clipped together, the number of these clips varying with the length of the spring and the use of which it will be subjected. Thus, Winton, Fig. 9, shows three clips and a band. Some springs show as many as five clips and two bands. None indicate the use of spring ears—very small projections on the ends of the leaves—which are bent over the edge of the leaf next below it to assist in holding the spring together, but they are in quite general use. Altogether, there are about 14 or 15 forms of spring-leaf ends, but those in general use may be reduced to seven. These are: the oval; the round point; the short French point, a modification of the oval; the round end with slot and bead; the ribbed form, widely used on motor trucks; the square point tapered; and the diamond point.

In addition, sizes have been standardized in America to the extent that only five widths are used for pleasure cars and seven for motor trucks. Those for the former are: $1\frac{1}{2}$, $1\frac{3}{4}$, 2, $2\frac{1}{4}$, and $2\frac{1}{2}$ inches; for the latter: 2, $2\frac{1}{4}$, $2\frac{1}{2}$, 3, $3\frac{1}{2}$, 4, and $4\frac{1}{2}$ inches.

As the automobile business has called for better stand-up qualities under more severe conditions of use, the quality of steel used has been greatly improved, and other materials are better. The French make excellent springs, many of our best automobile manufacturers going abroad for their springs for this reason, but American springs are improving in quality so rapidly that this is becoming unnecessary. Formerly, all springs were of a plain carbon stock, but now a great deal of silicon, manganese, and vanadium steel are being used. Some chrome and chrome-nickel steel have also been tried with considerable success.

SHOCK ABSORBERS

Function. The ordinary flat-leaf springs of any of the types previously described are inadequate for automobile suspensions. When the springs are made sufficiently stiff to carry the load properly over the small inequalities of ordinary roads, they are too stiff to respond readily to the larger bumps. The result is a shock, or jounce, to the passengers.

When the springs are made lighter and more flexible in order to minimize the larger shocks, the smaller ones have too large an influence, thus keeping the body and its passengers in motion all the time. These two contradictory conditions are the two conditions which have brought about the development and growth of the shock absorber.

The shock absorber is generally a form of auxiliary spring, the function of which is to absorb the larger shocks, leaving the main springs to carry the ordinary small recoils in the usual manner; in short, to lengthen the period of shock. This is done in a variety of ways, and, as it is only natural to expect, by a correspondingly large variety of devices.

General Classes of Absorbers. The simplest forms of absorbers are the ordinary bumper, or buffer, of rubber and the simple endless belt, or strap, encircling the axle and some part of the frame and acting as the rubber pad does—simply as buffer. There are the following classes of the more complicated shock-preventing and shock-absorbing devices: (1) frictional-plate or cam, in which the rotation of a pair of flat plates pressed together tightly—one attached to the frame, the other to the axle—opposes any quick movement of the two or of either one relative to the other; (2) a coil spring used alone and in combination—alone it is used in the plane of the coil, or at right angles to it, and parallel to the center line about which the coil is wound, while in combination it is found joined with the simple leather strap or with another coil spring of equal or sometimes of less strength, in the latter case the weaker one acting with the main springs; (3) the flat-leaf spring, a more simple description of which would be a small duplicate of the main semi-elliptic spring set on it so as to oppose its action; (4) the air cushion; and (5) the liquid device, in simple form and in combination with some one or more of the coil-spring forms.

Frictional=Plate Type. A frictional-plate type of shock absorber is shown in Fig. 23. This absorber consists of an upper arm attached to the frame, having at its outer end a frictional plate in contact with a similar plate at the upper and outer end of the other arm pivoted to

Fig. 23. Hartford Governed Friction Type of Shock Absorber
Courtesy of Hartford Suspension Company, Jersey City, New Jersey

the axle. The two plates are pressed together by means of the nut shown in the center; this nut is resisted by the spring beneath it and the slightly arched surfaces of the plates. When a sudden bump raises the axle, it must turn the two faces of metal across each other to the limit before it can lift the body. As will be seen, this means a considerable distance, and it can be made relatively greater by clamping the nut up tighter, thus increasing the friction between the surfaces, and, therefore, requiring greater force to turn them. Because of this adjustable quantity of friction, this type is called the governed friction type.

When cams are used, practically the same result is obtained, except that the device is necessarily more complicated. The cam action usually generates some heat, and, for this reason, this form of shock absorber is most always enclosed, and the interior, where the cam works, is filled with grease or very heavy oil.

A modification of the plain frictional-plate form is seen in Fig. 24, which is called a passive range absorber, because, for ordinary movements of the springs to which it is attached, it does not come into action. When the usual spring action is exceeded, however, as in a sharp jounce, the device becomes effective. It appears much like the Hartford just shown, but the construction is decidedly different. The upper, or frame, arm is threaded to receive an Acme-threaded screw, which is carried by the lower, or axle, arm. The action of screwing this out tends to force the plate on the lower arm, which must move outward with the screw against a rubber washer held firmly by the outside nut and cover plate. Thus, the scissors action of the two arms on a sudden movement is resisted by the compression of the rubber washer. This compression can be

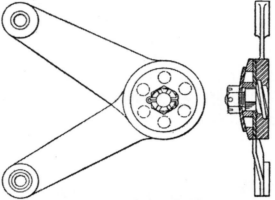

Fig. 24. Laporte Passive Range Friction Type of Shock Absorber

Courtesy of Charles Laporte, Detroit, Michigan

increased or decreased by tightening or loosening the slotted outside nut, so that the screw is given less or more movement. The rubber washer is made with a series of holes in it to allow of compression.

Coil Springs, Alone and in Combinations. *Springs Alone.* The coil-spring absorber is probably the most widely used form, primarily because it is both good and cheap; furthermore, it is simple and adds little weight. In most instances, the coil is so placed as to compress along the direction of its center line. One device, however, the Acme, Fig. 25, works at right angles to this. It consists of a pair of coils,

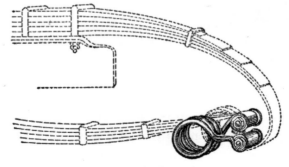

Fig. 25. Acme Torsion Spring Fitted to Three-Quarter Elliptic Gears

Courtesy of Acme Torsion Spring Company, Boston, Massachusetts

the two ends of each being so constructed as to go on the ends of the shackle bolts in place of the usual shackle. When the shackle is removed, one pair of ends is fastened to the spring in place of the shackle, while the other pair of ends is fixed to the frame or the other part of the spring, as the case may be. Note that this arrangement brings one of the coils on either side of the main spring end, extending away from it in a horizontal plane. In this position, the torsion spring acts as a spring shackle, a b s o r b i n g the jounces and bounces so that they do not reach either the body, the attaching point, or the other half of the spring, as the case may be.

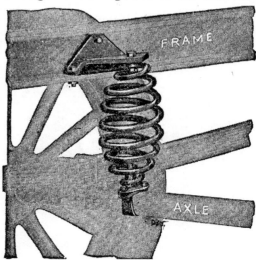

Fig. 26.   Sager Equalizing Springs Are Very Simple in Construction
Courtesy of J. H. Sager Company, Rochester, New York

Fig. 26, is a simple coil spring of barrel shape, that is, the end coils are smaller than those in the center and are set between frame and axle in such a way that they absorb the jounces directly. This is probably the simplest possible shock-preventing device, consisting only of the spring and its top frame and bottom axle connections. These are made in four sizes of wire, varying from $\frac{5}{16}$ inch up to $\frac{13}{32}$ inch.

In the K-W road smoother, shown in Fig. 27, the action of the spring is opposed by an air chamber at the top, creating a balance. A shock which causes the spring to move is opposed by the spring itself, while the rebound, or reaction, is opposed by the air compressed in the air chamber.

Combinations. Probably exceeded in simplicity only by the two forms just shown is the type in which a coil spring and leather band, or strap, are combined. One of these, the Hoover, is shown in Fig. 28. It will be seen that the spring end is fastened to the body, while the strap is

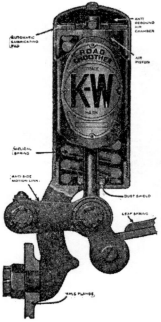

Fig. 27.   K-W Spring Type of Road Smoother

attached to the lower end of the spring and encircles the axle. Hence, this will not interfere with upward movements of the axle, but only with the downward ones, that is, the axle is free to rise, but as soon as the car body starts to rise, the strap-spring combination acts to prevent it. This is particularly true if the axle has reached the limit of its motion and has started downward before the body starts upward. In that case, the body can move upward only the amount of slack in the strap plus the give of the spring, but minus the amount the axle has already moved downward. This inexpensive arrangement has found great favor on small cars.

Fig. 28. Hoover Shock Absorber, a Spring and Strap Combination
Courtesy of H. W. Hoover Company, New Berlin, Ohio

Double-Coil Spring Types. In principle, the use of two springs is not different from the use of one. For structural reasons, however, it is easier to attach the two-spring form, while dividing the load up into two parts allows of the use of smaller diameters and smaller sizes of wire, thus making the device appear more compact. One of the two-spring forms, the J.H.S., is shown in Fig. 29. It consists of a pair of cylinders with coil springs within. The tops of the two cylinders are joined by a pin, and this joining pin is attached to the lower leaf of the spring. Inside the cylinders, pistons are set above each spring, and these are connected, this connection being used for the

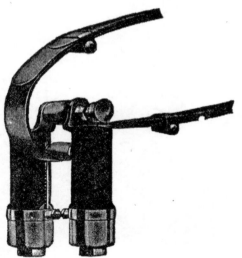

Fig. 29. J.H.S. Shock Absorber Has Twin Springs Encased

other half of the spring. At the bottom, the external bands on each of the two cylinders are connected, so as to keep them parallel at all times. Thus any movement upward of the lower part of the main-leaf spring tends to draw the enclosure for both shock-absorbing springs upward. The springs themselves resist this and absorb a large part of the movement both in force and distance.

Flat=Plate Recoil Springs. The third class, or flat-leaf spring, is a semi-elliptic unit in miniature. It is placed upon the top of the ordinary semi-elliptic spring, but it is reversed and has a spacing plate between the two. The object of this plate is to prevent recoil and to eliminate the rebound of the car body without restricting the flexibility of the main springs. As shown in Fig. 30, the Ames equalizing spring is constructed along these lines. As will be noted, this allows all downward movement of the spring, having no influence thereupon; but when the recoil, the upward equal and opposite reaction, comes, the smaller upper spring opposes this reaction and

Fig. 30. Ames Equalizing Spring Is a Simple Small Inverted Semi-Elliptic
Courtesy of Clarence N. Peacock and Company, New York City

minimizes it, so that little or none of it reaches the body or the passengers.

Air Cushion. Perhaps the most complicated form of shock absorber—certainly the most expensive and at the same time the most efficient—is the air cushion. This form consists of a pair of telescoping cylinders one being attached to the frame and the other to the spring. When road obstructions cause the spring to rise, it pushes its cylinder upward, but this movement is resisted by the air inside of the cylinders. With the amount of air properly proportioned to the size and weight of the car and its load, all this upward movement will be absorbed and none will reach the body and its occupants.

This rough outline describes the Westinghouse air spring, shown in cross-section in Fig. 31. In order to handle the air pressure and keep the cylinders within the commercial limits, oil also is used in the cylinders. This reduces the volume of contained air; but, for each inch the device is compressed, the air is reduced by a greater percentage of its original volume, consequently the resistance to compression is greater than it would be without the oil.

In the drawing, A is the upper section of the cushion chamber, telescoping into the lower section made up of tube B and crosshead E. The outer tube C is simply a guard. A steel casting D is bored out to form a guide for the outer tube and crosshead, and has a rectangular pad F machined for bolting the whole device to the bracket attached to the frame of the car. A shackle G is fastened to the end of the car spring I and is pivoted to the crosshead E. Packing ring H is used to make the inner cylinder a tight fit in the outer casing. A breather J is placed on the side, through which air is drawn by the upward movement of tube B through the medium of the tightness of packing ring H, just mentioned, and this air, on the downward movement, is forced through the passage K to a port partly surrounding the tube B. There is no packing ring between this tube and its guide D, so the air blows out and keeps the contacting surfaces clean. A further protection is afforded by the felt-wiper ring L,

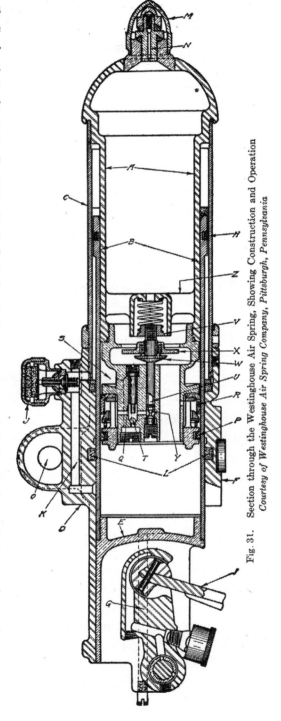

Fig. 31.   Section through the Westinghouse Air Spring, Showing Construction and Operation

Courtesy of Westinghouse Air Spring Company, Pittsburgh, Pennsylvania

which retains the grease in the groove just above it. O is a rod connecting the two front or rear springs. At the top is the screw cap M, covering the air valve N, which is designed to be used just as the air valve in a tire.

The lower part of the device is filled with oil up to a level which approximates the line Z, all above this level being air under pressure. Consequently, the device actually compresses the air through the medium of the oil, which is incompressible. This oil forms a seal for the air chamber and prevents its leakage, although the oil itself is allowed to leak through, this leakage being pumped back automatically by the action of the springs. This works out as follows:

Fig. 32.   Westinghouse Air Springs Applied to the Rear of Pierce Limousine

In what might be called the piston, although it is not, because it does not move—the other parts moving relative to it—there is the plain leather packing ring P and the cup leather R held out against the sides of the cylinder by the conical ring and spring.

The small amount of oil which does leak past the packing rings P and R is caught in the annular chamber S, whence it flows down through the vertical (dotted) passage Q into the chamber just below the ball valve T. In the center is a hollow plunger U of a single-acting pump. This has two collars on its upper end V and W and between them a disc X. This almost fills the passage just above it. The plunger is held down by the light spiral spring shown pressing on the collar V.

When a road obstruction is met and the spring rises, crosshead E rises and the upward movement of the oil takes the disc X upward until it strikes and carries with it collar V, which lifts the plunger and draws in a charge of oil. When the air compressed in the upper chamber of the device expands, and the car spring I and crosshead E go down again, the oil flows in the opposite direction, carries disc X down against collar W, and forces the plunger downward. Then the oil passes the ball check Y, goes through the hollow plunger, and is discharged back into the upper, or air, chamber. In the first place, the oil is put in by taking off cap M and taking out the air valve N. Then a special single-acting oil gun is used to force it in, a long nozzle being necessary to reach down into the interior, with a stop to limit this downward distance. The maker recommends that an excess be put in and then slowly drawn off to the right level.

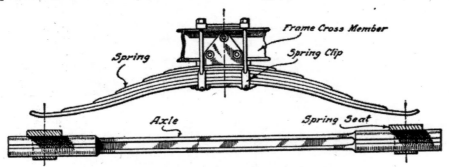

Fig. 33. Typical Semi-Elliptic Overload Spring

As will be seen from the foregoing, this device is essentially an air spring, and the air cushion does the work; but it is the oil below it, with its permissible leakage and with a pump to return this leaking oil, which makes this device practicable. To show the exterior, the part which most persons would see and remember, Fig. 32, is presented. This figure shows the rear end of a Pierce limousine equipped with a pair of the Westinghouse air springs. Note the breather, tie rod, cap at the top, cast guide at the bottom, and other parts previously shown and described.

Hydraulic Suspensions. The majority of the hydraulic devices developed as shock absorbers consist of turning vanes connected to the axle or spring, enclosed in a liquid-tight case filled with some heavy oil. There is a hole of small diameter in the case which connects the two sides of the vane, its motion forcing the fluid through this hole.

Thus the spring action simply pumps the coil from one side of the vane to the other and back again, the resistance to the flow of the liquid past the vanes and through the small hole absorbing all of the shocks.

Overload Springs. Overload springs are utilized with commercial vehicles and may be either the leaf or coil type, and so arranged as to act only when the load on the main springs reaches a certain weight. The wear plate may be a separate platform, as shown in Fig. 33, or it may be formed integral with the pressure block. Where coil springs are used, they are made of square section, attached either to the frame cross-member or to the axle. Two such springs are used, one on each side. The design in Fig. 33 is a semi-elliptic. It is attached to a frame cross-member, and the ends are free so that they may make connection with a separate spring seat or a pad on the pressure block of the side spring when a predetermined load has been applied. With some trucks the front springs are mounted on a seat forged integral with the axle and are retained by box clips; a coil spring is attached to the pressure block, which acts as a bumper. Under excessive deflections these springs strike the bottom flange of the frame and arrest the rebound motion of the vehicle spring. The Jeffery Quad employs a spring bumper which is made of flat metal and is termed a volute spring. It is attached to a bracket fastened to the pressure block.

SPRING AND SHOCK ABSORBER TROUBLES AND REMEDIES

Usual Spring Troubles. *Lubrication.* The average repair man is likely to have more call to lubricate the leaves of a spring than any other one thing in connection with springs. True, they lose their temper; they sag and show signs of losing their set; plates break in the middle, at the bolt hole, and near the ends of the top plate; and inside plates break in odd places. But more frequently the springs make an annoying noise, a perceptible squeak, because the plates have become dry and need lubricating. When this happens, and the up or down movement of the car rubs the plates over each other, dry metal is forcibly drawn over other dry metal with which it is held in close contact; naturally, a noise occurs.

To lubricate the spring, it is well to construct a spring-leaf spreader. Of course, the job is best done by jacking up the frame, dismounting the spring entirely, taking it apart and greasing each side of each plate thoroughly with a good graphite grease, then reassembling it, and putting it back under the car. This is the best way, but it costs the most, and few people will have it done. Sometimes spring inserts are used; these are thin sheets of metal of the width and length of the spring plates, having holes filled with lubricant over which is a porous membrane.

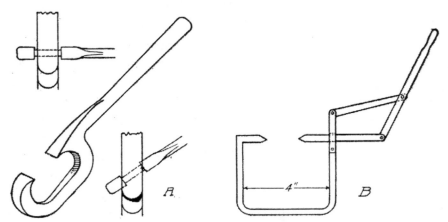

Fig. 34.   Handy Tools for Spreading Spring Leaves to Insert Lubricant

For the ordinary spreading job, the plates must be pried apart and the grease inserted with a thin blade of steel, for instance, a long-bladed knife. To spread the leaves, jack up the frame so as to take off the load, then insert a thin point and force it between a pair of leaves. In Fig. 34, two forms of tools for making this forcible separation are shown. The first is a solid one-piece forging with the edges hardened. It is used by sliding the edges over the ends of the spring leaf, then giving it a twist to force it in between them, as shown in the figures. The second tool is intended to be forced between two plates by drawing back on the handle.

Tempering or Resetting Springs. When springs lose their temper or require resetting, it is better for the average repair man to take them to a spring maker. Tempering springs is a difficult job, as it requires more than ordinary knowledge of springs, their manufacture, hardening, annealing, etc. When springs are in this condition, they sag down under load and have no resiliency. If a great many springs are handled, a rack like that shown in Fig. 35 is well worth making.

Broken Springs. When springs break, there is but one shop remedy—a new plate or plates. But when they break on the road, it is necessary to get home. When the top plate breaks near the shackled end, repair this sufficiently to get home by using a flat, wide bar with a hole in one end big enough to take the shackle bolt; bolt this bar to the spring in place of the end of the leaf which is broken.

Most of the broken springs are caused by the spring clips, that hold the axle to the springs, becoming loose. If the clips are allowed to remain loose, the spring leaves will shift out of place and will throw the axle and wheels out of alignment as well as breaking the spring center bolt which helps to keep the axle in the correct position.

It is an excellent plan to inspect the spring clips often especially after the springs have been removed for work to be done on them

Fig. 35. Simple and Well-Designed Spring Rack

because they settle down a certain amount and the clips become loose. Keep the spring clips tight at all times.

Spring Noises and Squeaks. During a certain amount of service, the sides of the spring and shackle plates become worn and allow a certain amount of side play of the spring on the spring bolt. This causes a sharp cracking sound and is very noticeable when the car is being driven over a rough road. The spring-bolt nut should be tightened on both bolts and, in some cases, it is necessary to insert thin pieces of metal between the shackle plate and spring. If, after all the play is taken up, at the points mentioned, the noise is not entirely eliminated, the trouble can usually be traced to a loose spring bracket on the frame. The rivets that hold the bracket to the frame become worn and allow the bracket to move. When tightening the spring bolts and nuts, they should not be pulled up too tight or the spring cannot move freely between the shackle plates. If the bolts are too tight, the shackle plates will hold the spring eye so tight that it will throw a greater strain on the spring master leaf and cause the spring to break off at the end, close to the spring eye, or the point at which the bolt passes through the spring.

Spring Squeaks. This trouble is usually caused by either dry spring leaves or dry spring bolts. In the first case, the leaves should be spread apart and well lubricated. In the second, it is often necessary to remove the spring bolt, clean and polish the belt thoroughly, cover it with grease and replace before the squeak can be cured.

Worn Shock=Absorber Bushings. A sharp cracking noise can often be traced to worn shock-absorber bushings and the only cure is to renew the bushings. Sometimes the stud that supports the shock absorber will work loose in the frame and this is another point at which a cracking noise will occur.

General Hints on Spring Repairs. As a rule, a break in a plate takes place where it does not prevent operating the vehicle, but it should be borne in mind that the damage to the plate subjects the other plates to extra work, and, unless the broken member be properly repaired or replaced, the others are likely to break. If one of the intermediate plates breaks in the center at the belt, tighten the spring clips as much as possible. Very frequently the rebound clips will be found to be loose, and missing clips also contribute to spring breakage.

The removal of a plate from or addition to a set is very likely to upset the grading of the construction. It is not practical to replace a broken plate with a new one because it is of the same width and thickness, but an expert spring maker should be called in to see that the set, or fit, is correct. The fitting of a leaf requires the services of an expert spring man; while it appears to be a simple matter, the lack of knowledge by some claiming to be spring experts is responsible for breakage after the spring has been repaired. The spring clips and the nut of the center bolt should be kept tight. The importance of preventing the accumulation of rust on the leaves and of lubrication has been commented upon.

When removing the springs to work on them, it is a very good plan to mark all leaves so that each one will go back to its original position. If the spring is carefully measured, it will be seen that one part of the spring is longer on one side of the spring center bolt than the other. This fact, if remembered, will often save the repair man a great deal of work because he is likely to get the spring in position with one end of the spring attached to the hanger before he notices that the other end of the spring is too short to be attached to the other hanger. To correct the trouble, the spring must be taken out and turned around and then attached to the hangers.

INSTALLING CYLINDERS ON CURTISS AIRPLANE ENGINE

FRONT AXLES AND BEARINGS

FRONT AXLES

Classification. Generally speaking, front axles may be divided into three classes: the Elliott, the so-called reversed Elliott, and the Lemoine. These typical forms of axles are themselves subject to further subdivisions. For example, there are many different forms of Elliott axles, each manufacturer having what is practically his own form. Again, the Lemoine, when used by other firms, has been built in a practically new form, taking the second maker's name. In this country, it is claimed that the axles made by Timken are sufficiently different from the Elliott and reversed Elliott, from which the principle was taken, as to deserve the name of Timken axles.

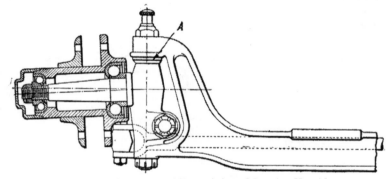

Fig. 1. Elliott Type of Front Axle and Steering Knuckle

Elliott Type. In general, a front axle consists of a bed, or axle center; a pivot pin or knuckle pin upon which the knuckles may turn; and the knuckles with the attachment for turning them. The Elliott type, Fig. 1, is the form in which the end of the axle takes a U-shape, is set horizontal and goes over the knuckles. The knuckles have plain vertical ends bored for the pivot pin, which passes through and has its bearing in the upper and lower halves of the axle jaw. In this form, the thrust comes at the top, where the axle representing the load rests upon the top of the knuckles that represent the point of support.

Reversed Elliott Type. In the reversed Elliott front axle, as the name would indicate, the action is just reversed in that the axle end forms a straight vertical cylindrical portion bored for the pivot pin, while the knuckles are so formed as to have jaw ends which go over the axle ends. The thrust comes at the bottom of the knuckle, where the axle bed rests upon the upper face of the lower jaw of the knuckle, the axle representing the load and the knuckle the support, just the reverse of the previous case.

This will, perhaps, be made clearer by illustrations. In Fig. 1, the axle has the jaw ends, and the thrust comes at the top. This is indicated in the figure by the letter A, which calls attention to the thrust washers at the top. Fig. 2 shows an axle of the reversed Elliott type, this being the front axle

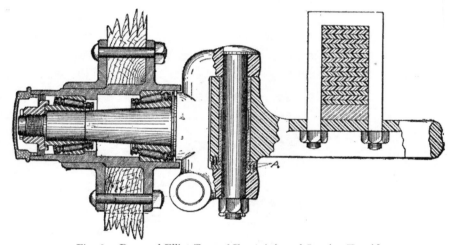

Fig. 2. Reversed Elliot Type of Front Axle and Steering Knuckle

for a heavy truck. In this the thrust washers A are at the bottom, and are of hardened steel, ground top and bottom to a true surface; the upper surface is doweled to the axle, while the lower is doweled to the knuckle. This form has the real advantage of concentrating all of the difficult machine work and assembling it into one piece, the knuckle. The Elliott type, on the contrary, makes the knuckle and axle difficult pieces to handle in the machine and afterward, this being shown in the cost. Ease of machining the bed of the axle is a great advantage, for the axle will average about 44 inches in length for a standard tread of $56\frac{1}{2}$ inches, and longer for wider treads, up to a maximum of about 48 inches for the wide-tread standard in the South.

The ordinary automobile machine shop is not fitted up for work of this size, particularly in machine tools other than lathes, and this job could not be done on a lathe. The result is that it becomes a task to handle it, necessitating special and expensive rigging for that one job. This was the case with the axle shown, a boring mill of the horizontal type and a large size milling machine being used on it. Both of these had to have special fixtures, which were useless at other times, to hold and machine these parts. At that, this job was much easier than an axle of corresponding size in the Elliott type would have been.

Lemoine Type. The Lemoine type of front axle differs from those described in that the axle proper bears upon the top or bottom of the knuckle-pin part of the knuckle, the two being made as one; that is, an extension or a jaw of the axle does not support the knuckle as with the Elliott type. When the steering knuckle of the Lemoine type is mounted below the axle stub, the latter is carried higher than with the reversed Elliott, so as to rest upon the top of the knuckle. An advantage of the construction from a manufacturing viewpoint is the cost of machining.

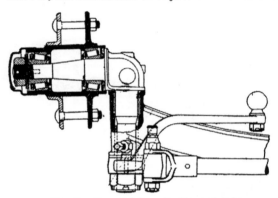

Fig. 3. Inverted Lemoine Type of Axle as Used on Overland Cars

With this design, the thrust load is practically entirely at the bottom upon the knuckle, which also must take all side loads; it is fastened in a sidewise direction at but one point—the bearing in the axle. The side shocks are taken on the end of a beam fixed only at the other end, whereas with the other types, the load is distributed between two supports, or divided equally over two sides, the point of support being midway between them. With the Lemoine type the bottom bearing must compensate for radial and thrust loads—a difficult condition to meet.

While the design is easy to machine, assemble, and handle, its disadvantage is that the knuckle has a double duty, having, as it does, both radial and thrust loads to care for because of its one-piece construction. This type of axle is, however, very popular with foreign designers.

Inverted Lemoine. A novel type of axle has recently been manufactured which is known as the inverted Lemoine. In this type, Fig. 3, the wheel spindle, or stub axle, is at the top of the steering knuckle instead of at the bottom as in the case of the regular Lemoine type. The knuckle has a single, fairly long support in the end of the I-beam front axle, the forging being much simpler on this account. In fact, this makes the axle nearly straight, which doubtless accounts in large part for this unusual design. One real advantage of this design is that it allows the car weight to be low in relation to wheel bearings, thus assisting in steering.

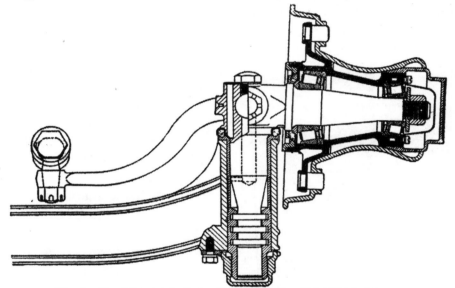

Fig. 4. Novel Front Axle Design Used on the New Light-Weight Marmon
Courtesy of Nordyke and Marmon Company, Indianapolis, Indiana

Marmon Self=Lubricating Axle. The new Marmon front axle, Fig. 4, is of the inverted Lemoine type similar to the Overland, shown in Fig. 3, but at first glance it looks quite different. For one thing, the bearing in the axle end is different, and in this lies an exclusive and valuable feature. The stub-axle pivot pin, made integral with the stub axle, is placed in a split bushing, which is a tight fit at the bottom—where the thrust collars are formed in it—and at the top, but not in the middle. When this bushing is in place, the knuckle and bushing are forced into the axle end from above, and a kind of hub cap screwed on at the bottom. This holds it permanently in place.

Near the middle of the split bushing there is a narrow slot to which a central bolt hole is connected. On being assembled, the inside is filled with lubricant, which cannot escape; but, as it wears away, the central bolt can be removed, more lubricant can be poured in until it is full, and the bolt replaced to prevent leakage. In this way the axle is self-lubricating, and, as the oil is used up very slowly, it needs practically no attention.

Like the inverted Lemoine, this arrangement of the axle end brings the axle down low, relative to the weight, and consequently steering is made easier. The lowering of the axle also brings the points of spring support down and thus lowers the whole car giving a lower center of gravity and a safer car in operation and reduces the weight.

MATERIALS

The materials utilized for front axles include castings of steel, manganese bronze, iron, and other metals, in the form of forgings, drop forgings, drawn or rolled shapes, and pressed shapes. Wood has been but little used and only in the past and then only for the lightest type of automobile.

Steel tubing is being used for front axles in the Franklin car and by other makers who are installing four-wheel brakes.

It is claimed that the front axle which is made of tubing will stand the torque and strain which is placed upon the axle at the time of using the brakes much better than the ordinary axle.

Cast Axles. Castings for front axles have been looked upon with grave doubt and fear by designers and owners, because of the fact that road shocks are more severe for front than for rear axles, and because of the fear that a casting may have a blowhole or some other defect. In addition to the natural distrust of castings for this work, it was feared that such material would crystallize more quickly than would a better and more homogeneous material like steel. There is, of course, a certain amount of crystallization in all materials, but far less in a close-grained fine-fibered structure like forged or rolled steel than in any form of casting. Aside from this, castings present many other advantages which are well worth while. Thus, the spring pads may be cast integral with the axle with practically no extra charge, while the same forged integral with a drop-forged axle may easily add several hundred dollars to the cost of the dies. Again, with casting patterns, the fillets may be changed easily to give a greater section here or to reduce a section there, while a similar action with any forged axle means a new set of dies, costing perhaps $600. There are many other machining helps which may be provided in cast axles without any extra cost.

Notwithstanding these many advantages, the casting for the front axle has been and is distrusted, and the makers who have used it have flown in the face of popular prejudice, for the public has mistrusted it even more than the makers. For this reason, the casting has been little used, and the writer fails to recall a single car with a cast axle now on the market.

Forgings. Forgings, as distinguished from drop forgings, are much used for good front axles, but are expensive. The writer knows of one excellent truck builder, striving to build the best truck in the world, who is using a hand-forged front axle, the end of which is shown in Fig. 2. It is forged down from a 6-inch bar of selected steel and the ends worked out so as to leave the bed proper a $2\frac{1}{2}$- by $2\frac{1}{2}$-inch section, which later has been increased to 3 inches square. This made a very costly piece of work, but the stand-up qualities shown in actual work more than made up for it as long as people could be found to pay the price demanded for a truck made along these lines.

Many smaller makers follow out the same scheme, the lighter work allowing the axles to be forged up much more quickly, more easily, and more cheaply. The smaller the amount of material to be heated, the less difficult will be the work, and the more quickly will progress be made. The general trend of axle practice today, however, is to turn over the axle job to specialists in that line, most of whom employ drop forgings, drawn- or rolled-steel tubing with drop-forged ends, or similar rapid-production forms of construction.

Drop Forgings. Drop forgings are now more used than any other form, although the first cost is great, for the dies must be very carefully worked out in a very high grade of steel; the result is a large expense of from $750 to $1200 before a single axle is turned out.

As a matter of fact, with drop forgings, after the die is once made, the axles may be turned out rapidly, accurately, and with little labor and cost. Given the dies, therefore, there is no doubt that this method produces an axle at a very low first cost. Moreover, the method itself produces better quality, for any process which works steel or wrought iron over and over again improves its quality, provided the steel is not burned in the process of heating. Not only are the majority of axles made of drop forgings, but of those not so made some part is almost sure to be a drop forging, as, for example, those made of steel tubing which have their ends or other parts made by the drop-forging process. In Fig. 5 is shown a drop-forged axle used on a truck.

Tubular Axles. The I-beam section of front axle is universally used, and while the tubular type formerly enjoyed some popularity, its use today is confined to a very few vehicles. When employed, its ends are drop forged or drawn, or rolled steel may be used with the ends welded or otherwise secured. The disadvantage of the tubular type is the fastening of the ends which is more or less offset by the lowered cost of material.

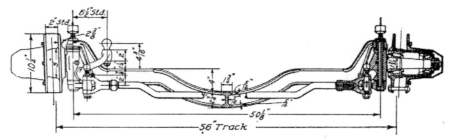

Fig. 5.   Typical Drop-Forged Axle Used on Truck

Drop-Forged Ends. Nearly all the ends for axles made in this way are drop forgings, very few castings being used, while the spring pads, or spring seats, as they are sometimes called, are split into upper and lower halves and bolted on.

The loading conditions of all front axles are such that the load rests on the axle at two points inside of the supporting points—the wheels. Thus, the continual tendency of the load acting downward and of road shocks acting upward is to bend the center of the axle still further downward. Since a tube which has been bent once has been weakened, it follows that this tendency to weaken it presents a further source of trouble.

Pressed-Steel Axles. The pressed-steel type of axle, which made its initial appearance in 1909, and is not generally employed, consisted of a pair of pressed-steel channel shapes—one being slightly larger than the other—set together with the flanges inward so as to present a box-like shape. When thus arranged, the two sections were riveted together by a series of rivets running vertically along the center part of the channels. The ends consist of drop forgings, machined to size or space between the channels when assembled, and then set into place between the ends and riveted. The pressed-steel construction obtained a secure attachment to the bed. This axle was of the Elliott reversed type.

Change of Axle Type Simplifies. Often the change from one type of axle to the other is not made because the latter is better but because of some incidental saving in the manufacture. Thus, in

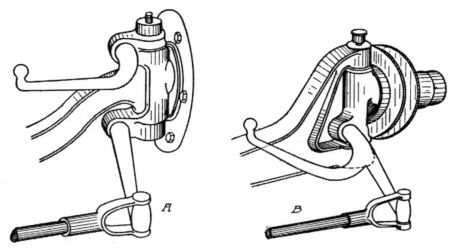

Fig. 6.   Differences in Construction of Reversed Elliott and Elliott Types of Axle Knuckle

Fig. 6, we see the reversed Elliott type at the left at A and the Elliott type at the right at B. From a manufacturing point of view, the former is much cheaper to construct, for the axle and knuckle costs would just balance one another, but the forging and machining of the one-piece steering arm shown in B would be more than double that shown in A. Moreover, the number of dies and their cost would be about three times as much, while the customer would have to be charged two or three times as much for repair parts. That is, in a modern low-priced car, produced in tremendous quantities, the advantages and costs connected with the two-piece steering arm of A would influence the choice of that design, regardless of other advantages or disadvantages.

FRONT AXLE TROUBLES AND REPAIRS

Alignment of Front Wheels Troublesome. The lack of alignment of front wheels gives as much trouble as anything else in the front unit. This lack not only makes steering difficult, inaccurate and uncertain, but it also influences tire wear to a tremendous extent. As Fig. 7 indicates, even if the rear axle should be true with the

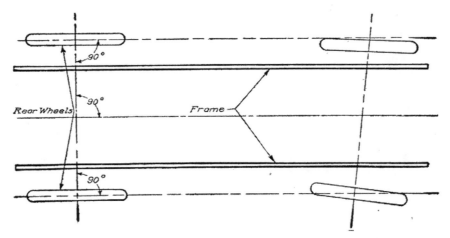

Fig. 7. Diagram Showing Front Axle and Wheels Out of True

frame, at right angles to the driving shaft, and correctly placed crosswise—correct in every particular with the shafts both straight so that the wheels must run true—the fronts may be out with respect to the frame, out of track with the rears, or out with respect to each other.

Fig. 8. Simple Measuring Rod for Truing-Up Wheels

In order to know about the front wheels, they should be measured; while this sounds simple, it is anything but that. In the first place there is little to measure from or with. A good starting place is the tires, and a simple measuring instrument is the one shown in Fig. 8. This instrument consists of a rod about $\frac{1}{4}$ inch in diameter and about 3 feet long, fitted into a piece of pipe about 2 feet long, with a square outer end on each, and a set screw to hold the measurements as obtained. By placing this rod between the opposite sides of the front tires, it can be ascertained whether these are parallel, and whether they converge or diverge toward the front. But knowing this, the driver or repair man is little better off than before, because this may or may not be the practice of the makers of the car, and it may or may not cause the trouble.

In short, a more accurate and more thorough measuring instrument is needed, Fig. 9. Such an instrument can be bought, but a

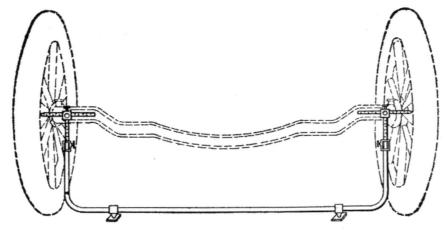

Fig. 9. Accurate Measuring Rod for Truing-Up Wheels. Better Design than Fig. 8.

similar outfit can be made from ⅜-inch bar stock, using thumb nuts where the two uprights join the base part, and also at the two points, or scribers, on these uprights. Having the floor to work from, the

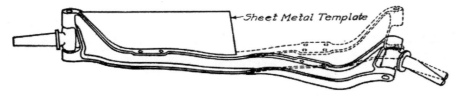

Fig. 10. Template for Showing if Axle Is Bent

heights can be measured, and thus the distance between tires may be taken on equal levels. Thus, a bent steering knuckle can be detected with this apparatus. Similarly, the center line and frame lines of the car can be projected to the floor, and by means of the instrument, it can be determined whether the axle is at a perfect right angle with the frame lines, and whether the wheels are perfectly parallel. Given the frame line, too, it can be determined whether the wheels track with one another.

Straightening an Axle. When an axle is bent, as in a collision, a template is useful in straightening it. This can be cut from a thin sheet of metal, light board, or heavy cardboard. It is an approximation at best and should be used with great care. Fig. 10 shows such a template applied to an axle which needs straightening.

When the axle is straightened to its original position, a pair of straightedges laid on top of the spring pads will be of great assistance in getting the springs parallel, as the worker can look across the straightedges with considerable accuracy. This is indicated in the first part of Fig. 11, which shows the general scheme. It shows also

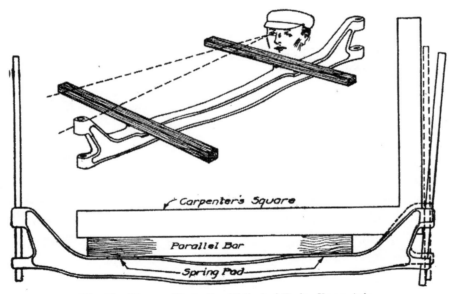

Fig. 11. Diagram Illustrating Method of Truing-Up an Axle

how the axle ends are aligned, using a large square on top of a parallel bar, but of course this cannot be done until the last thing, at least not until the spring pads are made parallel.

Front axles of light cars may be straightened without removal, provided the bend is not in the nature of a twist and not too short. Take two hardwood planks 7 feet long, 10 inches wide, and 2 inches thick. Cut four $\frac{3}{4}$-inch blocks 10 inches long and 3 inches wide.

Lay the blocks flat between the planks, space them about 2 feet apart, and bolt the whole securely. This obtains a girder 7 feet long, 10 inches wide, and $4\frac{3}{4}$ inches thick. Next, take two pieces of 4×4 timber 3 feet long and cut a tenon on one end of each. Make three $\frac{3}{4}$-inch eye bolts, 12 inches long, with nuts and plate washers for each. Place one of the eye bolts between each pair of blocks and screw up the nuts and washers sufficiently so as to rivet them. This permits of moving the eye bolts to any position between the blocks. The small steamboat ratchets and several short but strong chains complete the equipment.

With an axle bent back in the center, lay the girder on blocks in front of the car so it will be level with the axle, place the tenons of the 4×4 timbers in the space between the planks of the girder, one on either side of the bend, and connect the axle to the girder by means of a chain, the ratchet, and the eye bolt. When the ratchet is tightened up, it draws the ends of the 4×4's against the axle on either side of the bend. Tightening the ratchet still further removes the bend. This work may be accomplished in 20 minutes or less or in about one-tenth the time it will require to displace the axle, heat it, and straighten on an anvil, etc.

Spindle Troubles and Repairs. Wear of the spindle, or knuckle bolt, and its bushings, as well as play in the steering-gear linkage, brings about wobbling of the front wheels when the car is in motion. It is a simple matter to determine the component at fault. To test for bearing play, drive a block of wood between the knuckle and the axle, then grasp the wheel at the top and bottom, or at points diametrically opposite, and test for looseness. If none exists, the play is in the knuckle pin and its bushings. The remedy is to fit new bushings and new knuckle pins.

BEARINGS

Types of Bearings Required for Different Locations. As the portion of a mechanism upon which, more than upon any other element, its continued operation and long working life depends, the bearings of any piece of machinery should be of the most approved design and most perfect construction. The crankshaft and connecting-rod bearings, which are the most important on the motor, are of the plain type on the majority of engines. For camshafts there is more difference of opinion, while on fan shafts almost all makers use ball bearings. The other shafts, as pump, oiler, magneto, air-pump, generator, etc., are generally of the plain, solid, round type.

Engine bearings, however, are generally of the split, or halved, type, the upper and lower halves being practically duplicates. A reason for this construction appears as soon as one considers the application of the bearings to the shaft. It is granted that a crankshaft must be as firm and solid as possible, and hence it must be made in one piece. As ball bearings also are made in one piece, there arises at once the difficulty of getting the bearings into place on the one-piece shaft. This difficulty has necessitated cutting the shaft or else making it especially large and heavy in those cases where balls are used. With the split type of bearing there are no troubles of this kind and the bearings are adjustable for the inevitable wear.

Plain Bearings. The conditions that determine the proper proportioning and fitting of plain bearings have already been referred to in a preceding paragraph.

The materials of plain bearings are commonly varied to meet different conditions. With liberal bearing areas, in situations where it is desired to bring about a perfect fit with the minimum amount of labor, and to protect the shaft from wear in case there is failure of the lubrication, the various types of babbitt metal—which usually are alloys of tin and lead, with sometimes some admixture of antimony and other alloys—are widely regarded as the most serviceable. Probably the greatest advantage of a babbitted bearing is that, if the lubrication should fail, the low melting point and the soft material of the bearing will insure its fusing out without injury to the more expensive and valuable shaft.

Brass and bronze bearings, particularly the phosphor bronzes and the bronzes in which the proportion of tin is high and that of copper low, with sometimes the admixture of a proportion of zinc or nickel, will allow the use of materially higher pressures per square inch than can be safely permitted on babbitted bearings.

Steel shafts in cast-iron bushings, and even in hardened-steel bushings, make much better bearings than one might think, and though immediate trouble is to be anticipated with such a bearing should its lubrication fail, even momentarily, this trouble is more or less true of any bearing that can be devised. Since steel-to-steel and steel-to-cast-iron permit much the highest loadings per unit of area that are permissible with any type of metal-to-metal bearing, the merits of these materials are perhaps less appreciated than might be desirable. Steel pins through steel bushings, however, are not an uncommon construction for the piston-pin bearings in high-grade engines.

One noticeable feature of plain bronze or other plain bearings for automobile use is that they are always grooved for oil circulation. This is done by easing off the edges, then cutting a spiral groove by hand diagonally across to the other edge or to the center point where a similar groove from the other side is met. In a solid bearing, the groove is generally cut both ways from a centrally drilled oil hole, while in split bearings the grooves in each half usually form a modified letter x when viewed in plan, that is, two grooves start spirally inward from each edge near the ends, and all four meet in the center. This central point may be the spot where the oil enters or where it leaves. These grooves are seldom of very great depth, perhaps .008 to .010 (eight to ten thousandths).

New Oilless Bearings. A form of bearing that is new to the automobile but old in years is now coming into use. This is made of special wood which previously has been impregnated with oil. By saturating the pores of the wood with oil in this way, it is claimed that no lubricant need be used on the bearing for years. They are turned and fitted the same as bronze or other bushings.

Another oilless bearing is made of bronze with graphite inserts; this bearing is sufficiently soft to form the lubricant, yet sufficiently hard to retain its form and shape. Approximately one-half of the inner surface of the bearing is graphite, the two alternating in various ways, that is, the graphite is put in in spirals, in circles, in double diagonals, in a herringbone pattern, in zigzags, and otherwise.

Roller Bearings. Roller bearings, constituted by the interposition of a number of small rollers between shafts and casings, are a type of bearing widely employed in automobiles.

A much-favored construction is the tapered roller bearing illustrated in Fig. 12. This stands up very well under both thrust and radial loads.

Another type of roller bearing is that illustrated in Fig. 13, which is the type used on the Ford rear axle. This is one in which the rollers are small flexible coils made of strip steel, finally hardened and ground accurately to size. This type of roller can be depended upon to work without breakage or injury even though there be con-

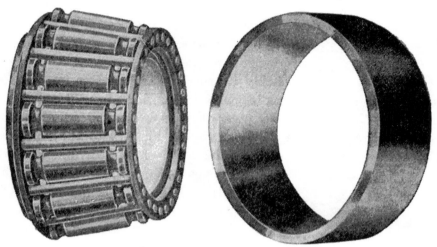

Fig. 12.   Timken Roller Bearing
Courtesy of Timken Roller Bearing Axle Company

siderable deflection or inaccuracy in the alignment of shaft or casings, the flexibility of the individual rollers taking care of such small errors.

Fig. 13.   Hyatt Flexible Roller Bearing Partly Disassembled to Show Components
Courtesy of Hyatt Roller Bearing Company, Newark, New Jersey

It will be noted in Fig. 13 that there is a solid steel shell to go on the shaft and fit it tightly, and another to fit into the case or support, whatever it may be, perhaps attached there permanently.   Between these two comes the cage carrying the flexible rollers. Any load imposed upon the shaft is transmitted to the inner sleeve and by it to the flexible rollers; these rollers absorb the load so that none of it reaches the outer case. Furthermore, shocks coming to the case from without are absorbed by the flexibility of the rollers and, *vice versa*, shocks to the shaft do not reach the case.

Ball Bearings. Probably the best of all bearings, except for certain special applications in which it is difficult to utilize them in sufficiently large sizes to assure durability, are the annular ball bearings of the general type illustrated in Figs. 14 to 18, inclusive. The basic feature of the most successful of modern annular ball bearings is their non-adjustability, the balls being ground very accurately to size and closely fitted between the inner and outer races so as to allow practically no play.

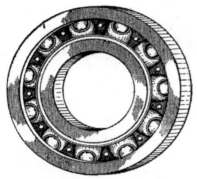

Fig. 14. Annular Ball Bearing

The reason that the best ball bearings are not made adjustable is that in any conceivable type of ball bearing one or the other of the races rotates and the other remains in a fixed position. The result is that there must be a loaded side to the race that does not rotate, with the consequence that when wear occurs, it wears the ball track deeper at this point than on the unloaded side. With the bearings thus worn, it is almost impossible to make an adjustment, for the attempt can result only in tight and loose positions as the balls come out and in of the spot that is more deeply worn.

Fig. 15. Section of Annular Ball Bearing

This condition has led the designers and manufacturers of the various types of high-grade annular ball bearings that are now on the market to discard adjustment as of no value and to substitute in its place qualities of material and hardness of surface which, in combination with the provision of sufficient sizes, are found to reduce wear to so small an amount that it is almost inappreciable. A bearing thus made can be therefore depended upon to outlive almost any other part of the mechanism in which it is placed.

The carrying capacities of ball bearings, as compared with those of roller bearings, are much greater than a casual consideration might lead one to suppose. Theoretically, the contact of a roller bearing—between a roller and one of the races—is a line contact, while that between a ball and a ball race is a point. But, practically, since some deformation occurs in even the hardest materials under sufficient load, the line contact in the roller bearing becomes a rectangle and the point contact in the ball bearing becomes a circle. Now the vital fact is that the area of the rectangle in the one case is substantially equal to that of the circle in the other—with given quality of materials and a given loading. So a ball bearing is fully as capable of carrying high loads as a roller bearing; besides, it avoids the risk of breakage

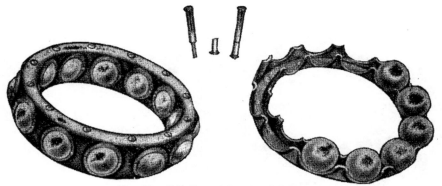

Fig. 16. Ball Cage of Annular Ball Bearing

that usually exists with rollers because of the impossibility of making them perfectly true and cylindrical.

To assemble ball bearings of the type illustrated in Fig. 15, either of two expedients may be adopted. One is to notch one or both of the ball races, so that by slightly springing them a full circle of balls can be introduced through the notch. The other scheme is to employ only enough balls to fill half of the space between the races, which permits them to be introduced without any forcing, after which they are simply spaced out at equal intervals and thus held by some sort of cage, or retainer, such as is illustrated in Fig. 16.

Ball bearings of the common annular type are quite serviceable to sustain end thrust as well as radial loads. For the best results under such loads, however, it is essential that the load be distributed equally around the entire circle of balls, for which reason the system illustrated in Fig. 17 is a means of avoiding the unequal distribution of pressure likely to result from the slightest inaccuracy of fitting.   In

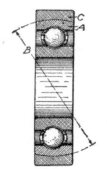

Fig. 17. Bearing Designed
to Equalize Loads

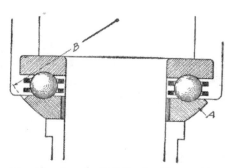

Fig. 18. Annular Ball Bearing Mounted for
Thrust Loads

this construction the outer ball race, shown at *A*, is provided with a spherical outer surface, permitting it to rock slightly in the mounting *C*, into the position shown in an exaggerated degree at *B*. It thus floats automatically to a position at exact right angles to the

Fig. 19. Bearing for Combined Radial and
Thrust Loads
*Courtesy of New Departure Manufacturing
Company, Bristol, Connecticut*

shaft upon which it is mounted, and so insures even loading of the whole ball circle.

An annular ball bearing designed for thrust loads alone is illustrated in Fig. 18. In this bearing, the lower race *A* is provided with a spherical face, described from the radius *B*, so that, as in the case of the bearing illustrated in Fig. 17, when in use it automatically floats under the load into such a position that all the balls are under equal pressure.

To secure uniformly satisfactory results from ball bearings, it is not only necessary in the first place to have them of the best materials, accurately made, and of sufficient sizes, but thereafter they must be always protected from dust and grit and from water and acids which tend to cause rust. They must also be kept lubricated.

Combined Radial and Thrust Bearing. The need for a bearing which would take ordinary radial loads well and also sustain thrust has led to the development of combined radial and thrust bearings, one being illustrated in Fig. 19. This is constructed to take either form of load equally well, and for this reason has displaced a pair of ball bearings in many circumstances where formerly it was thought necessary to use a radial ball bearing to sustain the load and a thrust ball bearing to absorb the end thrust. In this way it represents an important economy. Furthermore, it is economical of space, as it takes less room than the former pair of bearings used for the same two purposes.

When overhauling is done on any part of the car that contains a ball or roller bearing, the bearing should be carefully examined before it is put back into service. In the ball type, there should be no radial play and the races, cones, balls, and rollers should show no sign of pitting or cracking. If there is any sign of either, the bearing should be renewed because, if the bearing should fail in service, it is liable to cause a great amount of trouble. The bearing should be carefully washed in gasoline, thoroughly dried, and covered with a coating of vaseline, or grease. It should be wrapped in paper to prevent the dirt and grit from getting into the bearing. When the bearing is a tight fit in its housing, a soft punch should be used to drive it back into position and always driven in the ring that is the tightest fit. The blows should be equally distributed around the bearing ring. Be sure that the bearing is in squarely and not cocked so that the shaft, on which it fits, will be out of alignment.

AXLE BEARINGS

Classification. Thus far nothing specific has been said about axle bearings. These are, according to construction, of three kinds: plain, roller, and ball. From the standpoint of the duty which they are to perform, bearings may be divided into radial-load and thrust bearings, all three forms mentioned above being used for both purposes, but arranged differently on account of the difference in the work. Each one of the three classes may be further subdivided. Thus, plain bearings may be of bearing metal or of hardened steel, or they may even be so constructed as to be self-lubricating. Again, plain bearings may mean no bearings at all as in the old carriage days when the axle passed through a hole in the hubs, and whatever wear occurred was distributed over the inside of the hubs, resulting after a time in the necessity for either a new set of hubs or a new axle, or for the resetting of the axle, so that the hubs set further up on a taper. Roller bearings may be of several classes, some makers using both straight and tapered rollers. In addition to these there are combinations of the straight and tapered types, and bearings with two sets of tapered rollers acting back to back, the action being that

Fig. 20. Front Axle End, Showing Roller Bearings for Wheel and Steering Knuckle

of straight rollers, with the end-adjustment feature of the tapered type. There are also many types of ball bearings, as, for example, plain ball bearings—those working in flat races, those working in curved races, those working in V-grooved races, and single balls working alone. There are also combinations of balls in double rows.

Roller Bearings. Fig. 2 shows the use of tapered roller bearings for the hubs and of hardened-steel thrust washers for the thrust load, the figure showing, in addition, a plain brass bushing in the axle for the knuckle pin to turn in. In Fig. 20 is shown a more elaborate use of roller bearings of very excellent design. In addition to the axle bearing, it will be noted that the top bearing of the steering knuckle is of the roller type.

Ball Bearings.   Although there is a growing tendency to utilize a short adjustable type of roller bearing, many designers favor the ball

Fig. 21.   Front Axle and Steering Knuckle of Superior Construction

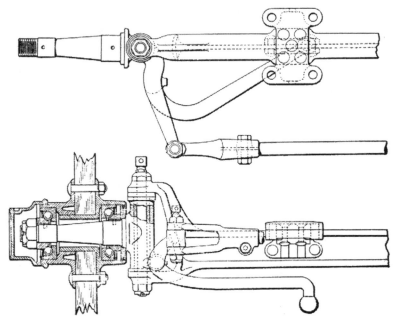

Fig. 22.   Front Axle Details of Waverley Electric Car
Courtesy of the Waverley Company, Indianapolis, Indiana

255

bearing. The two most common forms are the cup and cone type, which cares for radial and thrust loads, and the annular form which is suited for supporting annular loads. The annular form is not adjustable, and when it wears it must be replaced with a new bearing. The cup and cone type is adopted by makers of low-priced and medium-priced cars, has an angular contact, and is adjustable.

In some instances, particularly with high-grade cars, ball bearings are used for the knuckle bearings as well as for the hub. Fig. 21 is an example of an axle end, which for real bearing worth, has probably never been surpassed. The illustration shows the wheel hubs running on two very large diameter ball bearings, while the knuckle also turns on two very large ball bearings arranged for radial loads. At the top is another ball bearing arranged for thrust; this bearing taking up all thrust loads from the weight above or from road inequalities. Fig. 22 illustrates the cup and cone type. This design utilizes ball bearings for the hubs and plain steel thrust washers on the knuckle.

The only recent change in front axle design worthy of note is the use of the tubular axle which is used more extensively with four-wheel brake installations. The steering knuckle pins are also being given a greater angle which seems necessary with the use of four-wheel brakes.

FLINT SEVEN-PASSENGER SEDAN

WHEELS

There are three types of wheels used on the automobiles of today—the artillery wood wheel, as used on most pleasure cars; the steel spoke wheel, used on some pleasure cars; and the disc wheel, which consists of a steel disc attached to the hub and rim. The latter is rapidly gaining in favor. Commercial vehicle wheels are usually of the wooden spoke variety but there are a lot of steel spoke wheels in use on the heavier truck types. The material used in the manufacture of wooden wheels is well seasoned or kiln-dried hickory. The spokes and felloes of the wheels are made of this wood but there is now a tendency to use a steel felloe in which the spokes are set. The steel felloe is of the channel form and not solid.

Wheel Sizes. Wheels are used on automobiles, in combination with the tires, to afford a resilient and yielding contact with the surface of the road, so that people may ride with comfort. Therefore a wheel whose size is such as to yield the most comfort to the car occupants with due regard to its cost relative to the cost of the vehicle is the wheel to use. The cost of the wheels themselves, however, is so small in comparison with the cost of the pneumatic tires which are used on them as to be completely overshadowed by the latter.

Where comfort is sought as the prime requisite, cost becomes an accessory. The larger the wheel used the better the car will ride, and the greater will be the comfort of the occupants.

Large wheels were standard equipment in the past but the automobile of today is turning to the use of the smaller wheel for two primary reasons. The low car, with its low center of gravity, is more pleasing to look at as well as being a great deal safer to drive at high speeds than the high car. Large wheels raise the car from the ground, raising the center of gravity as well as destroying the pleasing appearance of the low "racy" car. There is also a limit to the size of the wheel that can be used when the gear ratio of the rear axle is considered. If a large wheel is used, the gear ratio will be increased and the power of the engine, as delivered to the rear wheels, will be reduced. This will be especially noticeable when hill climbing and acceleration are wanted. Another disadvantage of the large wheel is that it increases the leverage at the axle ends and, in turn, increases the strain that the axle must stand. The average sized wheel is about 34 inches in diameter. With the introduction of the balloon tire, the wheels are being made much smaller, in some cases as small as 20 inches.

At the present time, the larger tires are becoming more popular because of the increased riding comfort. It is not practical to use the large sized tires unless the wheels are altered to take the proper sized balloon tires. Unless this is done, the clearance between the tire and the fender will be insufficient and the under part of the fender will strike the tire when going over bumps and around corners.

In the case of the disc wheel, a special size can be obtained, but in the wood wheel, it is necessary to have a new wheel made that is suitable for the larger tires. The wheels that take the balloon tires are no different in formation except for a shorter and, in some cases, a thicker spoke. If the wheels are changed to take the balloon tires, it should be seen that the tires do not interfere with the operation of the brakes and that they do not rub against any projections as bolt heads and clevis pins.

It will often be found necessary when changing over from the ordinary wheels and tires to smaller wheels and balloon tires to place some small spacers behind the brake drums and brackets to prevent the interference of the tires with the brake operating mechanism.

If these changes are to be made it is best to take the car to the tire company and let them make the necessary changes and installation. This applies to the car that does not have balloon tires as standard equipment and is being changed over at the request of the owner, or customer.

PLEASURE=CAR WHEELS

Wood Wheels. Wood wheels are the most common form for pleasure cars in this country, being almost universal. Ordinarily, they are constructed of an even number of spokes, which are tapered at the hub end and rounded up to a small circular end with a shoulder at the rim, or felloe, end. Fig. 1 shows this construction, *A* being the felloe on which is the rim B, and R is the spoke which, at the hub end, tapers down to the wedge-shaped portion P. This matches up to the wedge-shaped ends of the other spokes, so that when the wheel is assembled they form a continuous rim around the central or hub hole.

The spokes are held at their inner ends by metal plates and by through bolts, which are set at the joints between the spokes so as to pass equally through each spoke, as shown at D. Not only do these bolts hold the spokes firmly to the wheel, but they have an expanding, or wedging, action tending to make the center of the wheel very rigid.

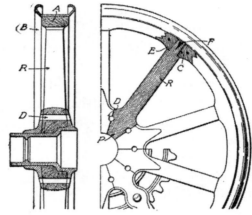

Fig. 1.   Construction of Wood Wheels

The outer end of the spoke has a shoulder E and a round part C, which fits into a hole bored through the felloe. To prevent the felloe coming off after the spoke is in place, the spoke is expanded by means of a small wedge driven into it from the outside, as shown at F. In this way, the wheel is constructed from a series of components into a strong rigid unit.

Such wheels wear in two places, at the inner and at the outer ends of the spokes. The remedy in the latter case is to withdraw the small wedge and insert a larger one in its place. At the hub end, when wear occurs, this, too, must be taken up by means of wedges. Fig. 2 shows

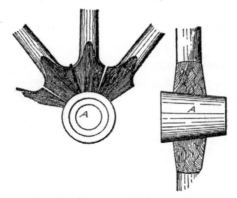

Fig. 2.   Method of Tightening Spokes of Wood Wheels

a method of doing this when the hub has no bolts at the joints. A false steel hub A is driven into the hub hole, after which wedges of steel are driven in between the wedge-shaped ends of the spokes. For slight cases of wear and squeaks, the wheel may be soaked in water, which will cause it to swell, taking up all of the space.

There are various modifications of this, nearly all of them changing the hub end of the spoke. In the Schwartz wheel, a patented form, each spoke is made with a tongue on one side of the wedge-shaped part and a groove on the other. In assembling the wheel, the tongue of each spoke fits into the groove of the spoke next to it, thus rendering the whole hub end of the wheel, when assembled, a stronger unit, being stronger in two directions, one of them of more than ordinary value. In driving the tongue into the groove, the wheel is rendered strong in a radial direction, but, when the wheel

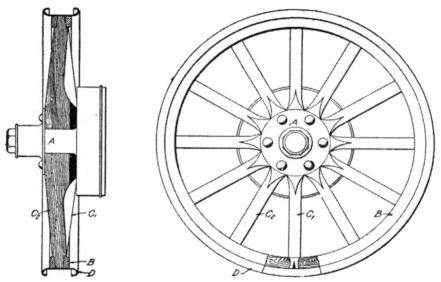

Fig. 3.   Details of Wood Wheels with Staggard Spokes

is entirely assembled, the tongue-and-groove method leaves it very strong to resist side shocks, a point in which the wood wheel is weakest.

Staggered Spokes. As mentioned above, the wood wheel has little lateral strength, nor can it ever have, from the very nature of its construction, except in unusual cases, like the Schwartz patent wheel just described. A method of increasing the lateral strength somewhat is that of using staggered spokes, these being alternately curved to the outside and to the inside, as shown in Fig. 3. This gives one set of half the spokes forming a very flat cone with its apex, or point, at the inner side of the hub, while the others form another cone with its apex at the outside of the hub. Each one of these conical shapes is stronger to resist stresses from the side on which the point is located than would be the same number of spokes set flat. Hence, the staggered-spoke wheel has the advantage over the ordinary type in that it has greater strength from both sides. In the figure, A is the iron hub, B the felloe, C_1 the right-hand and C_2 the left-hand spoke, and D the steel rim for the tire. This is a 12-spoke wheel, 6 of the right-hand spokes C_1 and 6 of the left-hand spokes C_2. The section shows how these pass alternately to the one side or to the other, forming the strong cone shape.

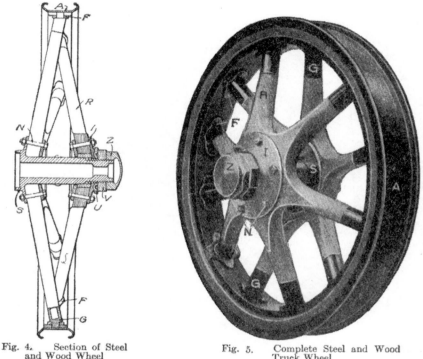

Fig. 4.   Section of Steel and Wood Wheel

Fig. 5.   Complete Steel and Wood Truck Wheel

Another method of handling this problem in a somewhat similar manner is the use of double sets of spokes, the spokes, however, being in two different planes separated a considerable distance at the hub. Of a necessity using the same felloe, the outer ends must be in the same plane. Fig. 4 shows a drawing representing a section through the center line of the wheel, while Fig. 5 shows a photographic reproduction of it.

In Figs. 4 and 5, A represents the steel rim on the felloe F, the latter being of metal in this case, as is also the wheel so it may be disassembled. The spokes R have a tubular end piece of metal G, which is set over the rounded end of each spoke and fits into a hole in the felloe. I and S are, respectively, the inner and outer parts of the hub, which are held together and to the spokes by means of the bolts N. Z is the hub cap, while U and V are filler pieces aiding in the dismantling process. The strength of the wheel is self-evident, but it is difficult to see the advantage of the disassembling feature, as a stress or strain which would break one spoke, would, in almost every case, break practically all of the spokes, thus necessitating a new wheel instead of new spokes.

Wire Wheels. Many of the little details of the automobile were inherited from its predecessor, the bicycle. Among these may be mentioned the wire wheel. Practically all bicycle wheels were

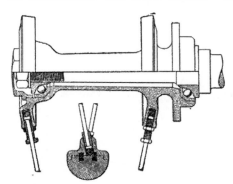

Fig. 6. Hub Details of Bicycle Wheel

and are of the wire-spoked type, and this same form of wheel was used on all earlier automobiles. It had no strength in a sidewise direction, nor did it, in fact, have much of anything to recommend it except its light weight. For this reason, it failed in automobile service, and received a setback from which it has even now not wholly recovered.

Early Bicycle Models. Fig. 6 shows an early type of wire wheel for automobiles, its construction indicating clearly its bicycle ancestry. The spokes were set into a casting, which formed the hub, and into the steel rim by means of a threaded sleeve, the head on each end of the spoke resting on the inner end of the sleeve. The sleeves were screwed in and out to adjust the tension of the spokes. This tension was usually considerable, thus reducing in part the ability of the wheel as a whole to resist side stresses, for the piece already in tension could not be expected to sustain additional tension, or compression, or a combination of either with torsion, according to the way the force was applied. Then, too, the casting for the hub was wholly unsuited to resist stresses, and the distance apart of the spokes at the hub was not sufficient, making the cone so very flat that it had very little more strength than a perfectly flat wheel.

Following the failure of wire wheels, there was a rapid change to wood wheels, which were almost universal for several years. Soon after this change was made, there was an increase in the size and power of automobiles, which, in turn, was followed by a demand for lessened weights. In the meantime, makers of wire wheels, knowing their faults, began to re-design in order to eliminate them. Their success is best evidenced abroad, where about one-half of the French and more than two-thirds of the English cars, in addition to over seven-eights of the racing cars in both countries, are now equipped with wire wheels.

New Successful Designs. This result has been brought about by a realization of the previous defects and their elimination. Thus, no more cast hubs are used, drawn or pressed steel of the highest quality and greatest strength being used instead. The spokes have been carried out farther apart at the hub, obtaining a higher cone and thus a stronger one. Spoke materials are

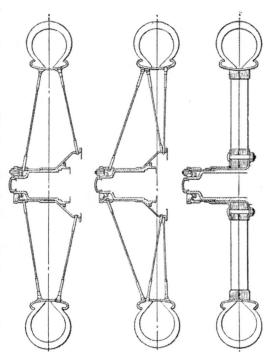

Fig. 7.   Sections of Double and Triple Steel and Wood Spoke Wheels

better and stronger, besides being used in greater quantities, that is, larger spokes and larger numbers of spokes per wheel, in some cases a triple row of spokes being used in addition to the ordinary two rows. This additional row acts as a strengthener and stiffener much like the diagonal stays on a bridge. Fig. 7 shows a set of double-spoke wire, triple-spoke wire, and interchangeable wood wheels side by side for comparison, while in Fig. 8 is presented a recent triple-spoke front wheel in detail.

In the former figure, the relative depths of the various cones and their corresponding strengths are made evident, being side by side. In this comparison, it will be noted that the new triple-spoke wheel has a much longer outer cone than the double-spoke wheel, while, on the other hand, the inner cone has been flattened. The triple spoke has a greater depth, considering the set of them as an additional cone, than has the inner cone in the double-set wheels.

In examining closely the older double-spoke form and the newer

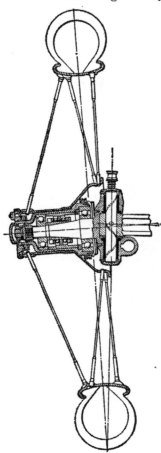

Fig. 8. Details of Triple-Spoke Front Wheel

triple type, it will be noted, also, how the wheel itself, or rather the tire and rim, have been brought closer in to the point of attachment, thus rendering the whole construction stronger and safer. In Fig. 7, it will be seen that the center line of both tire and rim passes midway between the inner and outer ends of the hub on double-spoke wheels, while on the triple form it is even with the inside end of the inner hub, being, in fact, farther in than is the case with the wood wheel. One thing will be noted in all these spokes, regardless of number, position, or inclination, and that is that their ends present a straight head. On the older bicycle spokes, the diagonal-spoke head was a great source of weakness, tending to create failure at the outset. The modern wire wheel is so constructed as to do away with this fault. By actual tests, the wire spoke—not the stronger triple spoke but the double spoke— has been found to have the following advantages: lighter weight for the same carrying capacity; greater carrying capacity for equal weight; superior strength from above or below in the plane of the wheel; lower first cost (it is doubtful if this will hold good for the newer triple-spoke forms); and, in addition, tests have proved superior strength in a direction at right angles to the plane of the wheel. So marked is the difference in weight of the two that five wire wheels are said to be lighter in weight than four wood wheels of equal carrying capacity.

All these arguments in favor of wire have been built up one by one, for much prejudice had to be removed. In spite of this, however, the wheel is slowly but surely building up a reputation and a long list of friends. Since, even now, England and the Continent continue to set the fashion in automobiles, it is not too much to expect to see wire-spoke wheels in common use in the United States in a few years. Most manufacturers to-day will fit wire wheels to their product for a small additional charge.

Wire Wheels Much Stronger. The increase in the use of wire wheels has been brought about by better designs; greater attention to the details of manufacture, assembly, and use; but primarily by the greater strength which has been built into the wire wheel. One way in which this has been done is by rearrangement of the spokes as, for instance, the triple-spoke form just described and shown in Fig. 8. Another and later form is the quadruple-spoke wheel as seen in Fig. 9. This is made and sold by the General Rim Company, Cleveland, Ohio, and is called the G-R-C wheel. As the sketch indicates, it has all the features of demountability, etc., of other wire wheels, the notable differences being the spoke arrangement to give strength and the form of rim—a patented form to be described in detail later.

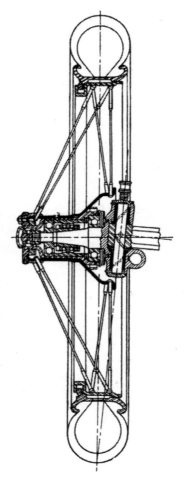

Fig. 9. G-R-C Quadruple-Spoke Wire Wheel

By comparison with Fig. 8, it will be noted that a double triangular section is formed in the G-R-C, the inner spokes forming the inside of the hub and the outside of the hub forming one triangle, while the outer spokes from each form the other.

In Fig. 8, it will be noted that there is but the one triangle and a straight row of spokes.

Wire Wheel Troubles. The spokes of a wire wheel should be inspected frequently as they work loose in service. This not only causes a creaking noise but will also cause the nipples of the spokes to work loose in the hub or rim and pull out of position which will cause the wheel to run out of true. Loose spokes also weaken the wheel and if allowed to get very bad might be the cause of the wheel collapsing. The wheel should be painted as soon as there is any sign of the paint chipping off especially around the nipples of the spokes as a prevention against rust. The wheels should also be kept running truly and the basis of this is to keep the spokes adjusted and tight.

Disc Wheels. Disc wheels are now being built with a single disc. Fig. 10 shows a full view and Fig. 11 a cross-section view of

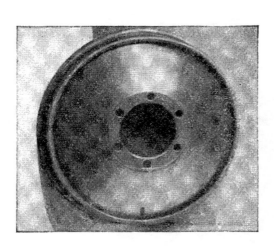

Fig. 10. Straight Disc
Courtesy of Budd Wheel Corporation

Fig. 11. Cross-Section
View of Straight
Disc Wheel
*Courtesy of Budd
Wheel Corporation*

a straight disc wheel. In the past, disc wheels consisted of two and sometimes three straight steel discs, Figs. 12 and 13, which were attached to the hub and the rim. There are still some trucks which have this type of wheel in operation, but solid tires are more generally used. The wheel used on the present-day automobile has a curved disc, Figs. 14 and 15, which gives a greater flexibility with

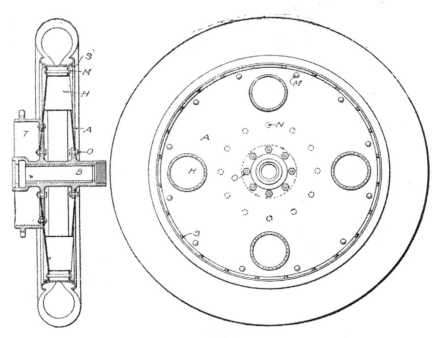

Fig. 12.   Side and Sectional Views of Sheet-Steel Wheels

Fig. 13.   Sheet-Steel Wheel Complete

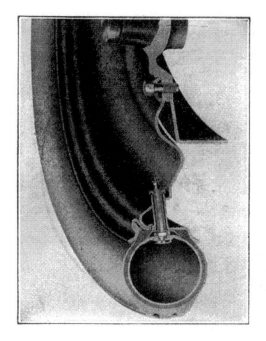

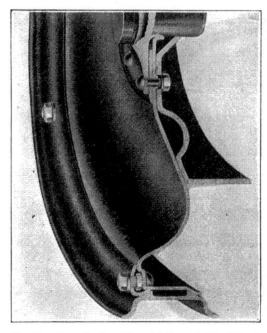

Fig. 14. Cross-Section Views of Curved Disc Wheels
Courtesy of Motor Wheel Corporation

strength and resiliency. The hub of the disc wheel is not a part of the wheel proper but is bolted to the hub and the rim is attached to the disc by riveting. The outer edge of the disc is formed into a flange and the rim riveted to it, the whole making a compact and strong installation.

There are two advantages in the use of the disc wheel. When tire changing is necessary, six bolt nuts around the hub are all that have to be taken off and the disc and tire is changed as a whole. If the car should hit the curb when skidding, the disc will not collapse as will the wooden spoke wheel. When the disc wheel first came into use, it was difficult to inflate the tire because the tire valve was behind the disc. Now however the valve projects through the disc. There is only one trouble worthy of notice in the disc wheel. At the point where the rim is attached to the disc, the rivets will sometimes work loose and cause a rattle. The rivets should be examined periodically and kept as tight as possible at all times. A peculiarity of the disc wheel is that it increases the noises developed in the running gear of the car. If the rear axle gears are inclined to be noisy or not correctly adjusted, the noise made by the gears will be greatly magnified because of the drum-like construction of the steel disc. The disc is also an asset to the car as they can be kept clean easily.

COMMERCIAL=CAR WHEELS

Requisites. On commercial cars the service is so different as to call for entirely different wheels. Of course, many commercial-

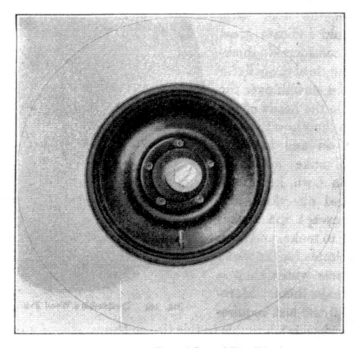

Fig. 15.   Front View of Curved Disc Wheel
Courtesy of Motor Wheel Corporation

car wheels are nothing but pleasure-car wheels with heavier parts throughout, but it is coming to be recognized that heavy trucks, tractors, and similar vehicles should have their wheels designed for the service required of them the same as lighter cars. No springiness or resiliency is required for heavy truck service, but simply these three things: strength to carry load and overloads; strength to resist side stresses; and such material, design, and construction as will make for low first cost and low cost of maintenance. A fourth desirable quality might be added to these, the quality of being adaptable or adapted to the tires to be used.

Wood Wheels. Taking Fig. 16 as an ordinary heavy vehicle wheel, let us see in what ways it fulfills or falls short of these requirements. The spokes are large in both directions and widened out at the felloe to give greater side strength. The felloe, which cannot be seen, may be judged as to size from the width and location of the dual tires, which would indicate great width and considerable thickness. This style of tire calls for a steel band shrunk over the felloe, while the heads of the cross-bolts show how the tires were put on and held on. All these make for great strength in both horizontal and vertical directions, and all parts except the spokes are simple to make, and even these are simple for the wheel manufacturer whose shop is rigged to make them. More-over, to fill the last require-

Fig. 16.   Double-Tire Wood Truck Wheel

ment, the wheel is adaptable to this tire or to any one of a number of motor-truck tires which might be used.

A slight variation from this is the double-spoke wheel, in which the spokes, in addition to being placed in double rows, are set so as to miss each other across the wheel, that is, each spoke of one row coming between two of the other. This placing allows the spokes to be made larger and stronger than in the ordinary case, while the double rows have the same strengthening effect as the tapering of spokes. The hub portion is assembled as two separate wheels, so that the work of assembling as well as of making the parts is slightly more than with the ordinary wheel. This is more than compensated for by the added strength. It is but fair to state that each of the last two wheels described is of English make.

In all wood wheels, the blocks composing the wheel and tire are of well-seasoned rock elm, sawed into wedge-shaped blocks, with the fiber lengthwise. The blocks are glued and nailed together until they form a circle. They are then turned round and to size in a large wood lathe, a shoulder $\frac{1}{2}$ inch wide being formed at the same time on each side of the tire 2.5 inches from the tire surface. A heavy steel ring with a corresponding shoulder is then shrunk

Fig. 17. White Cast-Steel Wheel

over the wood shoulder on each side of the tire, drawing it together much like the ordinary steel tire on a wood wheel of a carriage. Bolts are run through these rings and through the wood blocks from side to side to prevent the blocks from splitting sidewise. To increase the life of this tire, steel wedges $\frac{1}{4}$ inch thick are driven crosswise into the face of it 2.5 inches deep around the whole tire about 3 inches apart. These wedges prevent the tire from slipping; in fact, they act like an anti-skid chain and do not harm the pavement, being set flush with the surface of the wood blocks.

It is said that one set of these tires was used for nine months, and at the end of that time they were still good for service. The tires reach clear to the hub, thus doing away with spokes and enabling the tires to be slipped over the hub and held in place by a removable flange bolted through the wood to the fixed flange on the opposite side of the hub.

Cast=Steel Wheels. The heavier the service the more unsuitable do wood wheels become, that is, wood-spoke wheels. For many five-ton trucks, practically all seven- and ten-ton trucks, and nearly all tractors, the cast-steel wheel is used, either spoked or solid, the spoked form being given the preference. Fig. 17 illustrates a spoked cast-steel wheel, fitted with a solid tire. The wheel is cast with ten heavy ribbed spokes, a ribbed felloe, and a grooved-felloe surface, into which the tire is set.

Miscellaneous Wheel Types. *Steel.* Steel wheels are gaining for heavy truck use, and a number of the better steel-casting firms are now getting into this work, with the result that better steel wheels are becoming available.

Other constructions, such as steel and wood combination wheels with removable and replaceable spokes, and the like, are rapidly going out of existence. Truck work is unusually severe, and it takes but a few weeks of actual use to show up any of the so-called freak wheels. The simplest seems to be the best, the only question at present being whether the material shall be wood or cast steel. Pressed steel may offer some opportunities in combination with welding, since good work has been done on pleasure-car wheels of this type.

WHEEL TROUBLES AND REPAIRS

Wheel Pullers. In handling wheels a wheel puller of some form is generally a necessity; wheels are removed so seldom that they are likely to stick, and they get so much water and road dirt that there is good reason for expecting them to stick or to be rusted on. This means the application of force to remove the wheel. For this purpose, a wheel puller is needed, and a number of these have been illustrated and described previously, as gear pullers, steering-wheel pullers, etc. Any one of these devices which is large enough to grasp the spokes of the wheel and pull the latter outward and, at the same time, press firmly against the protruding axle shaft will do the work well.

Sometimes, however, while owning a puller, a wheel breaks down on the road where this is not available, or the repair man is called without being told the trouble, so that he does not bring the puller with him. In such cases, the repair man must improvise some kind of a puller out of what he has on hand. Everyone carries a jack, so it is safe to assume that one of these will be available as well as some form of chain. If a chain of large size is not available, tire chains—particularly extra cross-links—may be fastened together to answer the purpose. If chain is lacking, strong wire, wire cable, or, in a pinch, stout rope can be substituted. Attach the rope, wire, or chain to a pair of opposite spokes of the wheel,

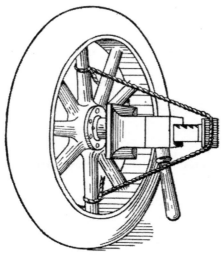

Fig. 18.   Makeshift Wheel Puller for Road Repair Work

Fig. 18, allowing usually about two feet of slack. Draw the chain out as tightly as possible, place the jack with its base against the end of the axle and work the head out by means of the lever until it comes against the chain. Then by continued but careful working of the jack, the wheel is pulled off the axle.

If rope, wire, or wire cable is used, it is advisable to place a heavy piece of cloth, burlap, or something similar over the head of the jack to prevent its edges cutting through this material.

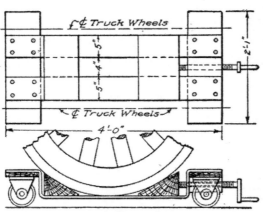

Fig. 19.   Tire Platform or "Dolly" for Handling Truck Wheels

With rope only enough slack must be used to allow the jack in its lowest position to be forced under it; this must be done because there is so much stretch to the rope itself and so little movement in the ordinary jack, that the combination of rope and jack does not always work to advantage.

Similarly, the handling of heavy truck wheels gives much trouble even in the garage, for they are so big, heavy, and bulky that ordinarily two men are needed. One man can do the trick, however, with a platform or "dolly" like that shown in Fig. 19. This consists of a platform about 4 feet long by 25 inches wide, fitted with casters at the four corners. Inside of the central part are placed a pair of wedges, one of which can be moved in or out by means of a crank handle. To use this, the wheel is jacked up a little over 2 inches, and the truck pushed under. Then the movable wedge is forced in against the tire so that the two wedges hold the wheel firmly and carry all of its weight. Then the casters are turned at right angles so that the platform and the wheel may be moved off together. The truck wheel is removed in the usual manner, that is, with the aid of the wheel puller or such other means as the garage equipment affords. The dolly also forms a convenient means of handling the wheel when it is put back on its axle.

Sometimes a wheel cannot be removed with a wheel puller and some other method must be used. With a bar behind the wheel, pry outward. Have an assistant give the axle a sharp blow which will jar the wheel loose. Place something soft on the end of the axle to take the blow, say, a piece of soft copper or lead.

Squeaky Wheels. A squeak in the wheel is usually caused by loose spokes and the trouble can sometimes be cured by tightening the wheel hub bolts. In other cases, it may be necessary to have the wheel thoroughly overhauled by an experienced wheelwright.

Worn Brake Drums on the Wheels. When a groove is worn in the wheel brake drum, it is necessary to remove the drum and face it down in a lathe. The wheel and drum should be marked so as to be sure that the wheel goes back to its original position.

GOODRICH BALLOON TIRE
Courtesy of Goodrich Tire Company

TIRES

Kinds of Tires. Broadly, there are three general classes of tires: the solid, the pneumatic, and the combination or cushion. The solid tire needs little comment or discussion here—being solely for commercial cars—except in so far as it is used with some form of spring wheel, hub, or rim. Similarly, the cushion tire is mostly used for electric cars, its use following that of the solid tire.

PNEUMATIC TIRES

The pneumatic tire was originally developed for bicycle use and in the beginning many single-tube tires were used. All of the tires used today have two parts—and inner and an outer tube.

Classification. Considering only the double-tube types, therefore, the pneumatic tire may be divided into three kinds: the Dunlop, the clincher, and various later forms brought out to go with the detachable demountable rims and similar devices. These latter vary widely in themselves, but all are modifications of the clincher form, with minor differences of the difference in rims.

Dunlop. The Dunlop tire, so named after the Irish physician who invented and constructed the first pneumatic tire, is brought down to meet the rim in two straight portions, perfectly plain and of even thickness, that is to say, the tire has no bead, as it is now called. The tire fabric is brought down to a straight edge at the rim, as well as the rubber covering, as shown in Fig. 1. *A* is the steel rim of the wheel, *B* the inner tube, *C* the outer shoe, which at the rim or inner portion is brought down to the two straight parts *DD*.

This tire, like all of the early tires, had to be put on over the edge of the rim by sheer strength, coupled with the flexibility of the tire when not inflated. This was a hard task, and, moreover, as soon as the tire was punctured or otherwise deflated, there was a strong possibility of its being thrown off, and possibly lost, at least after it had been stretched on and off the rim a few times.

Clincher. To prevent this latter happening, the clincher rim and tire were brought out, each being dependent upon the other.

In the clincher tire, the fabric is brought down to the **rim, and** then, instead of being left straight out as in the Dunlop, the material is formed into a hump, or bead, which is shaped just like the hollow

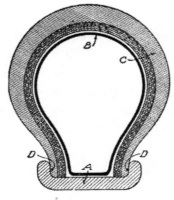

Fig. 1.   Section of
Dunlop Tire

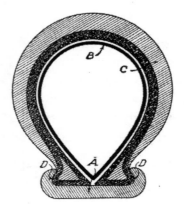

Fig. 2.   Section of Typical
Clincher Tire

formed in the rim.   The latter differs from the usual Dunlop rim only in having this deep depression to fit the bead of the tire.   Fig. 2 shows this, in which the parts are lettered as before.   In both

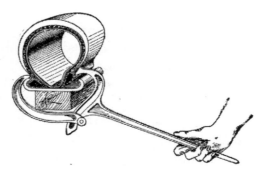

Fig. 3.   Tire Removing Tool

cases, the fabric of the tire is sketched in, and it may be noted that the layers are fewer in number in the older form.

The great majority of tires now in use are of this type, although, like the original Dunlop, it must be forced on and off the rim by the stretch of the deflated tire, and by sheer strength, coupled in this case with considerable natural ingenuity and some tools for lifting the hard non-stretchable beading over the edge of the rim at one point. This done, the rest is easy. For this purpose many tools have been bought; some good, some bad, and some indifferent. After a fashion, all do the work, but that tool is best which performs the operation most easily, most quickly, and with the least damage to the tire or rim. Fig. 3 shows a useful tool for this purpose.

The wire wheel and demountable rims, both allow quick road changes of damaged tires, leaving the work of tire repair to be done at home in the garage with proper heat, light, tools, and materials. This is rapidly bringing back into use the lower price clincher and straight-side tire forms, also many new tools have made their removal or attachment a much easier and more simple task.

Demountable Rim Types. Following the development of the clincher tire and rim until this form of tire was practically universal, came the first forms of the demountable rims, which consisted of a detachable edge on rim portion, like the edge of the clincher rim in section. These were locked in place in various ways in the different forms, but the first demountable rims—they were called detachable rims—were made by cutting the clincher rims into two parts, one of them detachable. This allowed of slipping the tire on over the rim in a sidewise direction, and did away with the stretching and pulling necessary with the plain clincher. Since this was a tire which was detachable more quickly than the ordinary tire, it was given the name "Quick Detachable," and now both parts are known to the trade as the Q.D. tire and rim.

Non-Skid Treads. All of the later developments in the clincher tire have been along the line of studded or formed treads to prevent skidding. In this many different things have been tried. Fig. 4 shows sections of many of the representative tires on the market. They are well known, and only the last three need any comment.

Fig. 4 *H* shows the Kempshall (English) tire tread, which is built up of a series of circular button-shaped depressions, or cups, which hold the pavement by means of the suction set up when they are firmly rolled down upon it. This tire has been very successful in England, but as yet had not been used much in this country.

The Dayton Airless tire, shown in Fig. 4 *I*, is a bridge-constructed cushion tire in which the usual air space is given over to a series of stiffening radial pieces of solid rubber, these with the tread forming the bridge or truss. Fig. 4 *J* shows the Woodworth adjustable tread for converting the usual smooth-tread tires of whatever shape or form into non-skids. It is a leather and canvas built-up structure, shaped like the exterior of a tire, and freely studded with steel rivets. When in place, the tire has all of the appearance of a leather-tread tire with steel studs.

Proper Tire Inflation Pressures. With the recent great increase in the value of rubber and the price of tires, the advice of manu-

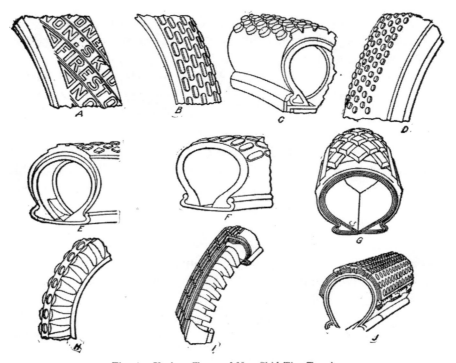

Fig. 4. Various Types of Non-Skid Tire Treads

facturers on the subject of tire wear is of great and growing importance. Nearly every manufacturer of tires is now recommending a table of inflation pressures which agree among themselves more or less closely. In each and every case, however, the makers are advising higher pressures than those generally used, stating that the people do not pump their tires up hard enough to get the best results from the materials in the tires. There should really be no conflict of interests here as the owner should be as anxious to get his mileage out of the tires as the makers are to make good their guarantees.

GASOLINE AUTOMOBILES

Many makers have stated, as a result of their years of experience, that more tires wholly or partially fail or wear out from under-inflation than from any other one cause. It thus behooves the owner of a car to look well to the pressure in his tires, not occasionally but very frequently. As the majority of gages attached to pumps in public garages are seriously in error, each motorist is advised to purchase his own gage—one of the pocket type which is simple and inexpensive—and carry it with him at all times.

In some cases, it will be found that pumping the tires up to the makers' specified pressure will result in unusually hard riding, and the motorist must be his own judge as to whether he wants to ride more comfortably and get less wear out of his tires or to put up with the discomfort and get every cent of wear out of them. In this matter, very few will choose the latter course.

Use of Standard Pressure and Oversize Tires. There is really a different way out. If the tire pressure advised by the maker results in too hard riding for comfort while comfortable pressure results in too much wear, the motorist is advised to get large size tires. These on the same car will have a greater carrying capacity than the weight of the car by a large margin. Just in the proportion of the tire capacity to the weight of the car will be the pressure recommended to the pressure utilized.

A simple example will make this clear: Suppose, for instance, a car weighing 3850 pounds, equipped with 34- by 4-inch tires, for which the makers claim a carrying capacity of 1100 pounds per wheel and recommend a pressure of 95 pounds. If this pressure be too high for comfort, and lower pressures, say 80 or 85 pounds, result in too rapid wear, the motorist should use larger tires. For instance, a 34- by $4\frac{1}{4}$-inch tire is scheduled to carry 1300 pounds per tire, and the pressure recommended is 100 pounds. The car weight per tire is 962 pounds, say 970. Changing to the larger tire gives a capacity of 1300 pounds per wheel, while the load is actually but 970. This change provides a surplus capacity which can be utilized to increase comfort.

Hence, if the tire be pumped up in the ratio of the carrying capacity of the tires to the actual weight carried, the spirit of the manufacturers' instructions will have been followed, comfort assured, and long life of the tire attained as well. Here the ratio of the capacity to the weight is as 1300:970. If now the pressure be figured from this, using the 110 pounds recommended, a suitable pressure will be obtained. Thus

$$1300 : 970 : : 100 : x$$
$$x \text{ —}74.6 \text{ pounds}$$

The pressure, therefore, in round numbers will be 75 pounds, and if this or any comfortable pressure above this be used, only the proper amount of tire wear will result, and a comfortable riding car will be assured.

However, this proposition, namely, changing from 34- by 4-inch to 34- by 4½-inch tires, is one which calls for entirely new rims, and possibly entirely new wheels, or at least new felloes, because the

Fig. 4. Balloon Tires—Showing Tire Flexibility
Courtesy of Firestone Tire Company

bottom diameter of the 34- by 4½-inch is different from that of the 34- by 4-inch. In such a case as this, the motorist would gain by changing to a still larger size, say 35- by 4½-inch, which change can be made without disturbing the old rims, as the 35- by 4½-inch is an oversize for 34- by 4-inch. This size also is recommended to carry 1300 pounds at 100 pounds pressure per square inch, but maximum pressure and comfort will be obtained from it between 72 and 80 pounds.

In general, the rule for oversize tires is this: Oversize tires are 1 inch larger in exterior diameter and $\frac{1}{2}$ inch greater in cross-section than the regular sizes, and any tire so sized will fit interchangeably with the regular size on the same rim.

Balloon Tires. The introduction of balloon tires seems to have met a popular public want and manufacturers are quickly adopting them as standard equipment. The use of balloon tires gives greater riding comfort because the air pressure is a great deal less than in

Fig. 5. Balloon and Standard Tire Comparison
Courtesy of Goodrich Tire Company

standard tires designed to support the same load. While the outside diameter is about the same, the increased tire size has been obtained by making the wheel, on which the balloon tire is used, smaller in diameter. It is claimed that increased mileage is obtained because the larger structure, giving greater flexibility, enables it to withstand the strains and distributes the wear over a greater area. These tires also give longer life to the car because the road shocks are absorbed by the air cushion, Fig. 4, and not transmitted to the car. Therefore the vibration is reduced, which is the greatest drawback to the long service of an automobile. The difference in size between the balloon tire and the standard tire of the same outside diameter is clearly shown in Fig. 5, and also by the cross-section,

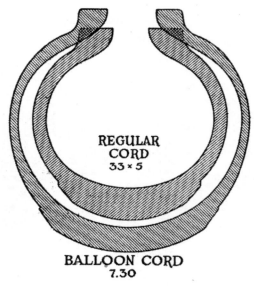

REGULAR
CORD
33 × 5

BALLOON CORD
7.30

Fig. 6.   Difference in Tire Sizes
Courtesy of Firestone Tire Company

Fig. 6.   There is also less tendency to skid with the balloon tire because the road contact area of the tire has been greatly increased. The difference between the standard, interchangeable, and balloon tires is clearly shown, Figs. 7, 8, 9, and 10. Useful information with regard to the sizes of balloon tires that replace the standard tire is also contained in these figures.

The flexibility of the balloon tire has been obtained by reducing the thickness of the side walls but because of the reduced air pressure used in these tires their efficiency has not been reduced one particle. If the standard tire was operated with a reduced air pressure, the thicker side walls would soon break down due to lessened flexibility. Some tire manufacturers claim that the interchangeable balloon tire can be safely used on the standard rim and that there is no difference in gear ratio, while others claim that the use of the super-oversized tire increases the gear ratio. The interchangeable balloon tire raises the car higher from the ground. To get the best results, however, all seem to agree that the wheel diameter should be reduced and the proper balloon tires fitted. In the fitting of these large tires, care

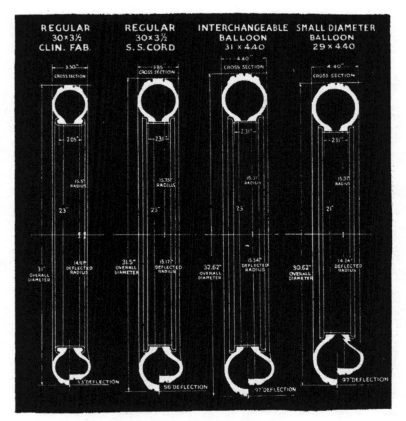

Fig. 7.   Interchangeable Tire Sizes
Courtesy of Goodyear Tire Company

should be taken to see that there is enough clearance between the tire and fender. Unless there is, the tire will bump against the fender when going over rough roads or around corners. Fig. 11 shows how to measure the distance between the tire and fender to check this clearance. Measure the distance at *A* from the center of the rim and the nearest point on the fender. Measure the distance at *B* from the rubber bumper to the nearest point of contact on the frame. The distance *A* should be greater than the distance *B*, as shown by the table below. While this table is especially suitable for

Goodyear tires, it will form a basis for measuring other tire distances.

31 x 4.40	5½"	33 x 5.77	7"
32 x 4.95	6"	34 x 5.77	7"
33 x 4.95	6"	35 x 5.77	7"
34 x 4.95	6"	35 x 6.60	8"

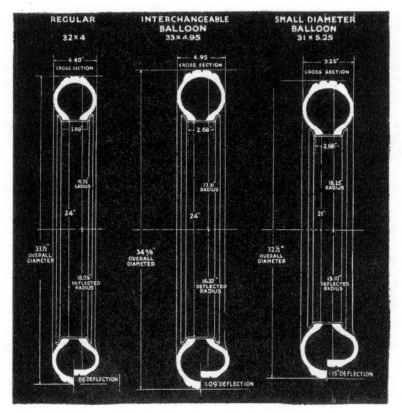

Fig. 8.   Interchangeable Tire Sizes
Courtesy of Goodyear Tire Company

The installation of tires on standard wheels is not in fitting balloon tires but in installing super-oversized tires. No specific air pressures can be given for all balloon tires as each maker has his own idea as to the air pressure suitable for their tires. The mechanic, or owner, should be sure to get the correct pressure from the makers for the make of balloon tire installed.

GASOLINE AUTOMOBILES

TIRE CONSTRUCTION

Composition and Manufacture. Tires consist of two parts: The tube and the shoe, or casing. The former is a plain ring of circular cross-section, made of pure rubber, containing an air valve, and is intended only to hold the air. The shoe, or casing, on the other

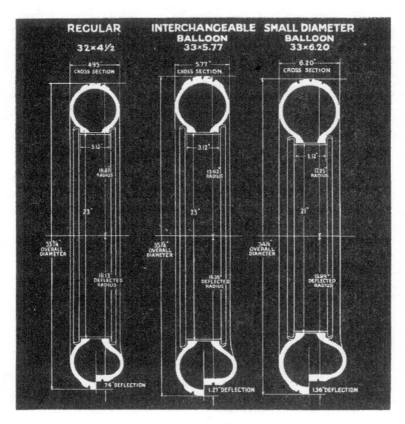

Fig. 9. Interchangeable Tire Sizes
Courtesy of Goodyear Tire Company

hand, provides the wearing surface, protects the air container within from all road and other injuries, and constitutes or incorporates the method of fastening itself to the wheel. In its construction are included fabric—preferably cotton—some pure rubber, and much rubber composition, the whole being baked into a complete unit by heat in the presence of sulphur, which acts somewhat as a flux for rubber.

Considering a typical tire, there enters into its make-up, starting from the inside, six or seven strips of frictional fabric, that is, thin sheets of pure gum rubber rolled into intimate contact with each side of the cotton, making it really a rubber-coated material. Next,

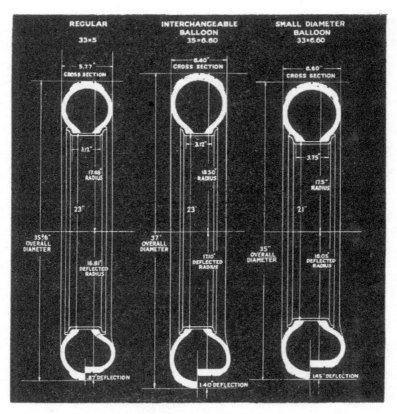

Fig. 10.   Interchangeable Tire Sizes
Courtesy of Goodyear Tire Company

there is the so-called padding, which is more or less pure rubber, has a maximum thickness at the center of the tread, and tapers off to nothing at the sides, but usually carrying down to the beading. Above this there is placed a breaker strip, consisting of two or three layers of frictioned fabric impregnated in a rubber composition. This, too, is thickest at the center and tapers off to the sides, but ends at the edge of the tread. Finally, there is the surface covering, called by rubber men the tread; this contains very little pure rubber, being thickest at the center and extending with gradually decreased thickness almost down to the bead.

The last two of this series of layers constitute the real wearing surface of the tire, and when the surface is so worn that the breaker strip may be seen, it is time to have the tire retreaded. When the

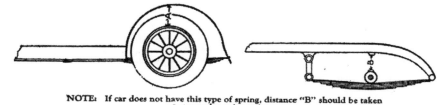

NOTE: If car does not have this type of spring, distance "B" should be taken between bumping points.

Fig. 11.   How to Measure Tire Clearance
Courtesy of the Goodyear Tire Company

wear has gone through this, if the padding be fairly complete, retreading will still save the tire, but if wear has gone clear down through that so as to expose the fabric, the tire must be run to a finish and then discarded.

All this construction can be noted in Fig. 12, which shows a section through a tire, with the inner tube in place, the section being taken so as to pass through the center of the tire valve. This should be borne in mind when examining this figure, for the location of the inner tube inside the tire, as previously described, is likely to be misleading.

Bead. In the reference to tire construction, no mention has been made of the bead. This is a highly important part of the tire, for it is the part which holds it in place on the wheel. It is made of a fairly hard rubber composition, the fabric being

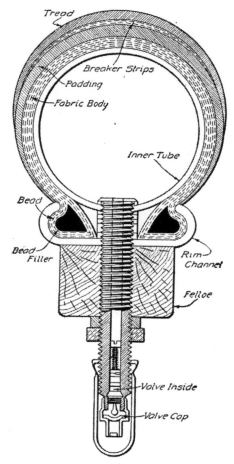

Fig 12.   Section through Assembled Tire and Tube, Showing Construction and Parts of the Tire and Tire Valve

carried down on the sides so as to cover it. In a cross-section, it has a shape very close to an equilateral triangle resting on its base; around the wheel it is curved to fit the rim. The method of attaching the tire has a considerable influence on bead construction, since, in the clincher type of tire, in which the shoe must be stretched on over the rim, the bead must be extensible in order to insure easy mounting. In the quick-detachable and straight-side forms of tire there is no need for this stretching, so the bead can be made of stiff and rigid material as well as cut down somewhat in size.

The straight-side or Dunlop type of tire is seldom made with much of any bead, the layers of fabric being carried straight down.

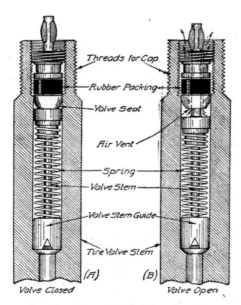

Fig. 13. Views of Tire Valve, Showing Closed and Open Positions

A more modern form of tire has a pair of woven-wire cables incorporated in the bead to make it stiffer and stronger, and this is said to have been very successful. As has been pointed out previously, this could be done only with the quick-detachable form, not with the clincher type.

In both the clincher and the quick-detachable forms, the bead holds the tire to the wheel by means of parts of the rim, which bear on it from above, as well as sidewise, the internal pressure when the tire is inflated pressing it against these parts very firmly.

Tire Valves. In Fig. 12 there is shown a section through the tire valve but on a small scale. As this is a very important part and little understood, a larger view is shown in Fig. 13. This is in two parts, *A* at the left showing the valve closed, and *B* at the right indicating the position of the various parts when the valve is open. The lower part of the valve is hollow, so that air inside of the tire has access to the valve seat. Note that the valve is held down on this by the threaded portion above it. This valve seat forms a slight taper which rests against an equally slight taper inside of the valve stem.

One condition of the tire valve holding air pressure is that the two valve seats be clean and smooth and free from scratches or cuts and foreign matter. Now it will be observed that the valve-seat portion of the valve has a hole through the center, in which the stem is a loose fit. This large hole passes all the way up through the threaded portion. The stem has a projection below the valve seat, which normally is held up against the bottom of the seat by the spring, this being strong enough to hold it up so tightly that no air can pass between the two. There are other conditions for valve tightness. The spring must be strong enough to hold these parts together; and the surfaces must be clean and true so that when held together, no air can get through.

Action of Valve. The action of the valve is this: When air is pumped in, it passes down around the central stem until it meets the projection, which it forces down against the pressure of the spring and, when there is air inside, against the pressure of the internal air. As soon as this is pressed down, the air passes in, and if the external pressure is stopped, as at the end of a stroke of the pump, the spring and the internal pressure push the projection back into the place, and no air can escape. On the next pressure stroke of the pump, this is repeated, the whole process continuing until the tire is filled.

Leaky Valves. It will be noted that with a good clean spring, projection, and valve seat, the pressure of the air itself holds the valve tight. Thus, when a valve leaks, it is a sure sign that some part or parts of it are not in good condition. If the valve is not screwed down far enough, air can leak out around the valve seat, so that leakage may be remedied by screwing the whole valve farther down into the stem. If the valve stem is too tight a fit in the central hole, it may stick in a position which allows air to pass. This can be remedied by a drop of oil placed on the stem and allowed to run down it. But not more than one drop should be used as oil is the greatest enemy of rubber, and the tube with which the valve communicates is nearly pure rubber.

If the spring is too weak to hold the projection against the bottom of the valve seat, the valve will leak. This can be remedied by taking out and cleaning the spring, also stretching it as much as possible. In general, however, the best plan of action with a troublesome tire valve is to screw it out and put in a new one. These can be bought for fifty cents a dozen, and every motorist should carry a dozen in a sealed envelope, also a combination valve tool. When trouble arises with the valve, or a tire leaks down flat with no apparent cause, screw out the valve with the tool, screw in a new one, make sure it is down tight, and pump up again. The few cents it will cost to throw away a valve, even if it should happen to be good, will be more than compensated for by the time saved.

Washing tires often is a good practice, since water does them no harm, while all road and car oils and greases will be cleaned off, nearly all of these being injurious. Frequent washing will also serve to call the attention of the owner to minor defects while they are still small enough to be easily repaired, and thus they are prevented from spreading. When not in use, tires should be wrapped, so as to be covered from the light, and put away in a dry room in which the temperature is fairly constant the year round. They will not stand much sunlight, nor many changes in temperature. Cold hardens the tires and causes the rubber to crack. Heat has a somewhat similar effect and also draws out its life and spring.

In general, of all things to be cared for and repaired promptly, no one thing is of more importance than the tires. If this rule is kept in mind, better satisfaction in the use of the car will result. So, too, with other repair work; if tools and appliances are made available and repairs made as soon as needed, the car will be better understood and give more satisfaction than if the opposite course is pursued. A few months of use of a car will do more to emphasize this than any amount of talk.

Inner Tubes. Improvement has been made in inner tubes by the use of better and purer rubber in much thicker sections. Some of these have a partial fabric reinforcement; others are made and then turned inside out so that the tread portion is under compression, thus resisting punctures or internal pressure. Other designs present a tube larger than the inside of the tire before inflation; this produces a truss formation, which the air pressure stiffens.

Cord Tires. The real improvement of value, however, is the cord tire. One form of this is shown in partial section in Fig. 14. This shows graphically that the difference between this tire and other forms is that the 4 to 6 or more layers of fabric have been replaced by two layers of diagonally woven cord. This cord is continuous, rubber impregnated, rubber covered, and, through its size, allows a great and very even tension. Lessening the amount and thickness of the fabric has given a greater percentage of rubber in the tire; consequently, the cord tire is more resilient. The advantages claimed for it are: less power used in tire friction, which means more power available for speed and hill climbing; greater carrying capacity in same size; saving of fuel; greater mileage per gallon of

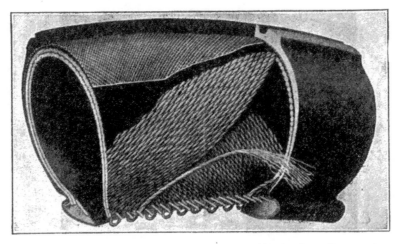

Fig. 14.   Section of Goodrich Silvertown Cord Tire, Showing Inner Construction

fuel; additional speed; quicker starting; easier steering, thus less driving fatigue; greater coasting ability; increased strength; and practical immunity from stone bruises owing to superior resiliency.

Another type of inner tube is one that is made on a form so that the tread side has a greater circumference than the side nearest the wheel rim. This is a preventive against stretching the tread surface which would make it easier to puncture the tire; also prevents kinking the rim side.

TIRE REPAIRS

Repair Equipment

Vulcanization of Tires for Repair Man. In practically all of the following material the point of view is that of the professional repair man, or of the garage man about to take up tire repairs, as distinguished from that of the average owner or amateur repairer. The lesser tire injuries and their repairs are handled from an amateur standpoint in another part of this work.

Vulcanization, to the unitiated, sounds very mysterious, but it really is nothing more or less than cooking, or curing, raw gum

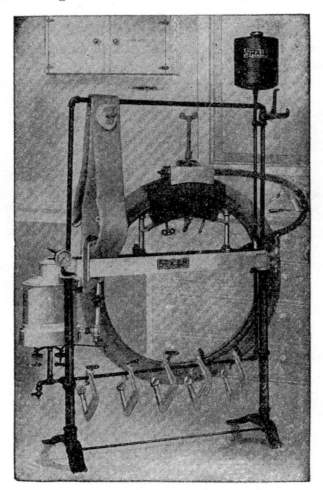

Fig. 15. Small Vulcanizing Outfit for Single Casing of Six Inner Tubes
Courtesy of C. A. Shaler Company, Waupun, Wisconsin

rubber. In the processes of manufacture a tire is cooked, or cured, all the component parts supposedly being united into one complete whole. A tire is repaired preferably with raw gum or fabric prepared with raw gum, and, in order to unite this to the tire, vulcanization or curing is necessary. The curing, in addition to uniting the parts properly, gives the proper strength, or wear-resisting qualities, which raw rubber lacks.

Types of Vulcanizing Outfits. *Shaler Vulcanizer.* This curing, or cooking, is done by the application of heat, in a variety of ways. Generally, very small individual vulcanizers have a gasoline or alcohol cavity, holding just enough of the liquid so that when lighted and burned the correct temperature will be reached and held for the correct length of time. The larger units are operated by steam or electricity; the latter is preferred for its convenience, but the former is used by the majority of repair men. The source of heat is immaterial so long as the correct temperature is reached and maintained for the right lengh of time. Too hot a vulcanizer will burn the rubber, while too low a temperature will not give a complete cure.

For the average small repair man, the outfit shown in Fig. 15 will do very nicely, at least to start with. This will handle a single casing or six tubes, or in a press of work, both simultaneously. This outfit is operated by gasoline, contained in the tank shown above at the right, but the same outfit can be had with pipe arrangements for connecting to a steam main, or for electric heating. In the case of either gasoline or steam, there is an automatic temperature controlling device which is a feature of the Shaler apparatus. As shown, casings are repaired by what is known as the "wrapped tread method", the repair being heated from both inside and outside at once, the outside being wrapped. Tubes are handled on the flat plate, shown in the middle of the framework, the size of which is $4\frac{1}{2}$ by 30 inches, this being sufficient, so the makers say, to handle six tubes at once.

Haywood Vulcanizer. For larger work, a machine something like the Haywood Master, shown in Fig. 16, is excellent. This is a self-contained unit, carrying its own gasoline tank, steam generator, and other parts. It handles four casings at once, while the tube plate G, 5 by 18 inches, is large enough for from three to four tubes, according to the allowance per tube made in the Shaler outfit. The separate vulcanizers are not designed for the same part of a casing, a side wall and bead vulcanizer being shown at D, a sectional vulcanizer for large sizes at E, a sectional vulcanizer for small and medium sizes at F, and a side wall and bead vulcanizer for both clincher and straight-side tires at H. The gasoline tank is marked C, with vertical pipe in which is the gasoline cut-off valve K. This leads down to the gasoline burner M, where the gasoline in burning vaporizes the water into steam. The water gage L, which indicates the amount of water available, is placed on the side of the steam generator A. Above this steam generator is the steam dome at B,

Fig. 16.   Master Vulcanizer with Self-Contained Steam Generator
Courtesy of Haywood Tire and Equipment Company, Indianapolis, Indiana

from which the steam pipes lead to the various molds. The returns, or rather drips, will be noted, also the steam gage (not marked) and the cut-off valve in the supply pipe to the sectional molds. In addition to the molds shown and a full supply of parts and tools, sectional vulcanizers for $2\frac{1}{2}$- and 3-inch tires, relining mold for $2\frac{1}{2}$-, 3-, and 3½-inch tires, and relining mold for 4-, 4½-, 5-. and 5½-inch casings come with the device.

This outfit with the extra molds, described but not shown, gives a very complete equipment for the small shop doing average

Fig. 17.   Battery of Vulcanizing Mold for Various Sizes of Tires

repairing.   In fact, when a shop outgrows this type of equipment, it must specialize in tire work and purchase special equipment.

Separate Casing Molds for Patch Work.   In the way of separate molds for casings, an excellent example of the localized heat type is shown in Fig. 17.   By this is meant the form designed to vulcanize a small short section of a tire.   The illustration shows five sections capable of handling, respectively, 2¼-, to 3-inch (motor-cycle), 2½- to 3-inch (small car), 3½- to 4-inch, 4½- to 5-inch, and 5½- to 6-inch tires, thus covering the entire range.   These molds have a special arrangement in that the heating portion is divided into three sections, into each of which steam can be admitted separately.   This allows the use of one, two, or all the sections, according to the nature of the repair.

In Fig. 18 is shown how it is possible, with this apparatus, to vulcanize the tread portion only by admitting steam solely to the larger bottom steam chamber around the tread, similarly, with the right-hand bead or side wall or the left-hand bead or side wall.   When a complete section is to be vulcanized, all sections are opened.   The importance of this

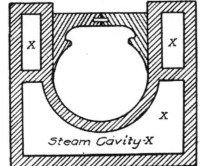

Fig. 18.   Section of Vulcanizer, Showing Steam Cavities

will be realized in a simple consideration of the fact that the tire itself has already been vulcanized and further heat is not only not good for it, but is distinctly bad, as it deteriorates the rubber.   Where the heat is needed, however, is not the raw rubber which has just been added at the repair point, this being practically useless until it has been cured.

Vulcanizing Kettles. *Horizontal Type.* When it comes to vulcanizing an entire tire, as, for instance, when a new tread has been

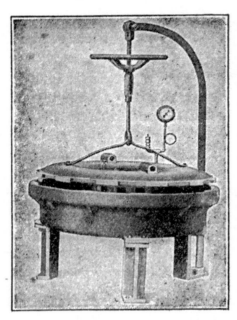

Fig. 19. Vulcanizing Kettle, Horizontal Type

put on, or other very large repair, what is known in the trade as a "kettle" is needed. This is simply a heavy steel tank, large enough to take one or more entire tires, steam being admitted to its interior to vulcanize them. The kettle shown in Fig. 19 has a capacity of two casings 36 inches in diameter or smaller. It is of the type in which no bolts or nuts are used for fastening the cover, this being held fast by the projecting lugs which lock under other projections on the top of the kettle when the cover is turned. A special rubber packing ring also is used, Fig. 20, effectually sealing the kettle against steam leakage. This kettle resembles a doughnut in shape, the tires lying within the circular cavity.

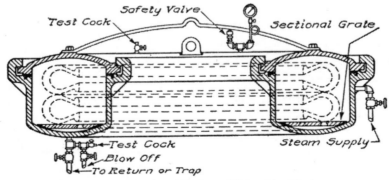

Fig. 20. Section of Horizontal Vulcanizing Kettle

Large Vertical Type. When the work goes beyond the capacity of size and type of tank or kettle shown in Fig. 19, which will handle two casings at a time, and at least two, perhaps four, kettles full an hour, that is, from 40 to 75 casings a day, it becomes necessary to use a larger type of kettle, made in vertical types only. These consist simply of large round steel shells with hinged heads, into

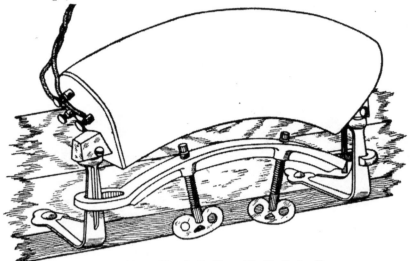

Fig. 21. Shaler Electrically Heated Inside Casing Form

which the tires can be rolled and piled, after which steam is admitted to the whole interior. They vary in size from 36 inches inside diameter by 24 inches in length to 48 inches diameter by 40 inches in length.

Inside Casing Forms. Another requisite of the tire specialist is an inside casing form, such as is shown in Fig. 21, or something similar. Many tire repairs are inside work, and even on those which are external, it is important to have an inside form against which the tire can be pressed and firmly held while vulcanizing. This particular form is heated by electricity, the wires being shown at the left; it is 14 inches long and has an external shape to fit the inside of all casings.

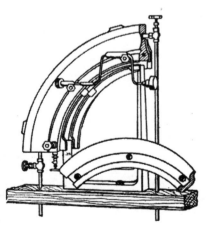

Fig. 22. Side-Wall Vulcanizer

Side=Wall Vulcanizer. A shop doing a great deal of work can use to good advantage the side-wall vulcanizer shown in Fig. 22.

It has a single central member through which the steam passes, and also has bolted-on side plates, the insides of which are formed to suit either clincher or straight-side tires. In the figure, the side plates are not both in place, one being shown on the work table below. The brace shown is used to remove the clamping nuts quickly and easily. This form is very useful on all side-wall or bead operations. It applies **greater** pressure along these parts of the tire than an air bag; it exactly

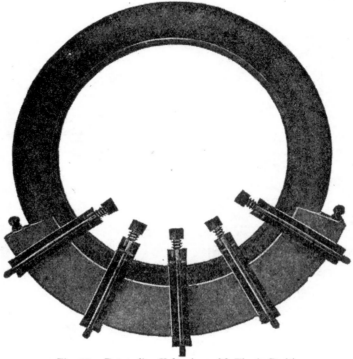

Fig. 23. Retreading Vulcanizer with Tire in Position
Courtesy of Haywood Tire and Equipment Company, Indianapolis, Indiana

fits the tire, and the size and shape make it possible to vulcanize a 36-inch tire in four settings.

Retreading Vulcanizers. Retreading vulcanizers differ from the sectional molds of Figs. 15, 16, and 17 in that the heat is applied at one particular point or, rather, strip along the middle of the top surface of the casing and extending down only as far as the side walls. Such a device, shown in elevation in Fig. 23, and in enlarged sectional detail in Fig. 24, is used solely for retreading or vulcanizing a new tread strip around the tire. The complete unit extends around about one-third of the whole tire surface so that when putting on a complete new tread the mold must be used **three** times. The section, Fig. 24, is numbered as follows: casing, 2; inner mold, 1; new tread to be vulcanized, 3; vulcanizer proper, 4; clamp, 5; and steam space within which the heating is done, 6.

Layouts of Equipment. There are two ways of installing **an** outfit somewhat like that just described, namely, by the non-return system and by the gravity-return system.

Non-Return Layout. A typical installation according to the non-return system is shown in Fig. 25. A steam trap must be placed in the system to remove the water and discharge it either into the sewer or into a tank so that it can be used again. In the figure there is shown a tube plate, a three-cavity sectional vulcanizer, two inside molds, and a medium size kettle of the vertical type placed in order from right to left. A pressure-reducing valve is shown which permits the use of a higher pressure in the boiler, thus maintaining an even steady pressure on the vulcanizers regardless of fluctuations at the boiler.

Fig. 24. Section of Retarding Vulcanizer
Courtesy of Haywood Tire and Equipment Company, Indianapolis, Indiana

Gravity-Return Layout. When the coil steam-generator or flash type of boiler is used, the gravity-return system is utilized, this being a method of piping by means of which the condensed steam is returned to the coil heater to be used over again. This makes it necessary to set the apparatus so that the water of condensation will run back to the coil heater, which means that the pieces must be in a series, each successive one being set a little lower down to the boiler. Figs. 26 and 27 show a side view and plan view, respectively, of a small

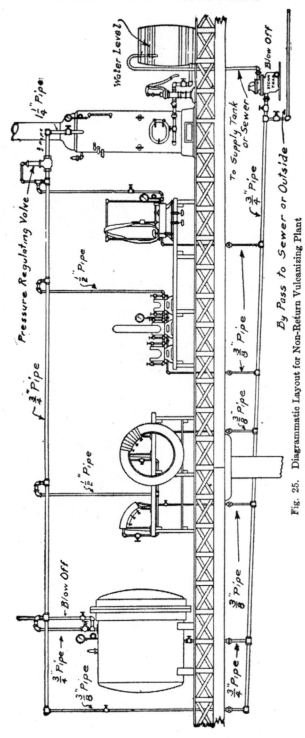

Fig. 25. Diagrammatic Layout for Non-Return Vulcanizing Plant

plant arranged on this plan. The outfit consists of the coil heater, which may be fitted to burn gas or gasoline, two inside molds, a large tube plate, and a three-cavity sectional vulcanizer. The outfit

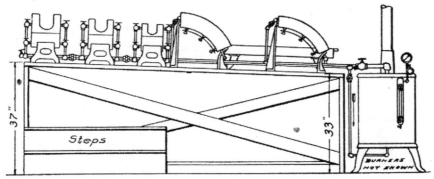

Fig. 26. Elevation of Gravity-Return Vulcanizing Plant

differs from Fig. 26 only in the absence of the kettle; on the other hand, the tube plate in Fig. 26 is larger.

Small Tool Equipment. In addition to these larger units, the well equipped tire repair shop should have a considerable quantity of small tools, among the necessities being those shown in Fig. 28. At *A* is shown a flat hand roller and at *B* a concave roller. *C* shows an awl, or probe, which is used for opening air bubbles and sand blisters. *D* is a smooth stitcher; *F* a rubber knife, of which two sizes are advisable, a large and a small; and *G* a 10-inch pair of shears for

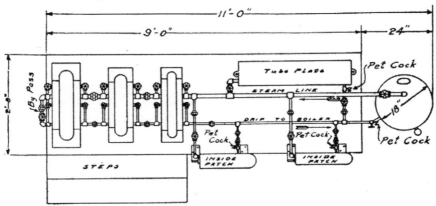

Fig. 27. Plan View of Gravity-Return Vulcanizing Plant

trimming inner tube holes, cutting sheet rubber, etc. *H* is a steel wire brush for roughing casings by hand; a preferable form is a rotary steel wire type driven by power at high speed. *I* is a similar wire brush for roughing tubes; and *J* another brush with longer wires, also for roughing casings; *K* is a tread gage for marking casings to be retreaded; and *L* a fabric knife necessary in stepping down plies of fabric. *M* is a pair of plug pliers for placing patches inside of small tube repairs; *N* is a cement brush for heavy casing cement, another very much smaller and lighter one—preferably of the camel's hair type—being used for tube cement. *O* is a hand

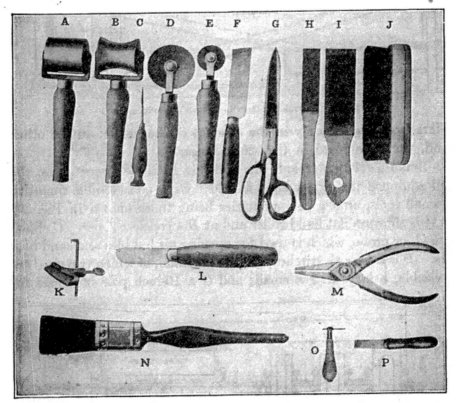

Fig. 28.   Collection of Tools Necessary for Vulcanizing Work

scraper and *P* a tread chisel; *Q* performs a somewhat similar function, being a casing scraper for cleaning the inside of a casing preparatory to mending a blowout.

In addition to the small tools shown in Fig. 28, it is necessary to have several tube-splicing mandrels; a large number of various sizes and shapes of clamps for all purposes; rules, try-squares and other measuring tools; tweezers for handling small patches, tools for recutting threads on tire valves; tire spreaders, for holding casings open when working inside; a casing mandrel or tire last of cast iron for holding a casing when making repairs; a tread roller for rolling down layers of raw stock evenly and quickly; a considerable amount of binding tape; thermometers; and such motor-driven brushes, scrapers, etc., as the quantity and quality of the work warrant.

Materials. Each repair shop must carry such a supply of tire-repairing material as the nature and quantity of its business demands. Among other things may be mentioned: Tread stock, rebuilding fabric, single-friction fabric, cushion stock, breaker strips, single-cure tube stock, combination stock, cement, quick-cure cement, soapstone, valve bases, valve insides, valve caps, complete valves, vulcanizing acid, various tube sections, tire tape, cementless patches, as well as many other tire accessories to sell. Many good tire-repair shops find a legitimate use for special tire-repairing preparations on the order of Tire-Doh.

Inner Tube Repairs

In general, all tire repairs come under one or more of the following headings; puncture; blowouts; partial rim cut or rim cut all around; and retreading or recovering, and relining.

Simple Patches. Under the heading of punctures are handled all small holes, cuts, pinched tubes, or minor injuries. Generally, these can be repaired by putting on a patch by means of cement, or with cement and acid curing. When well done, this method is effective. This kind of a job seldom comes to the repair man, and, when it does, it is principally because the owner is too lazy to do the work. About the only two cautions necessary are relative to cleanliness and thoroughness. The tube and patch should be thoroughly cleaned. Again the patch should be large, well cemented, and the cement allowed to dry until just sticky enough to adhere properly. Many a simple patch of this kind has been known to last as long as the balance of the tube.

Large Patches. *Cleaning the Hole.* Whenever the hole or cut is large, it is recommended that the repair be given more serious attention and vulcanized. The ragged edges of the rubber should be trimmed smooth with the tube shears or knife, the minimum amount of rubber being cut away. The hole, however, should be made large enough to allow the insertion of an inside patch. Then the tube around the hole should be cleaned thoroughly. This is best done with a cloth wet with gasoline, cleaning not only the outside but the inside around the hole and at the edges. In order to make a good job of this, it should be gone over several times; the larger the hole the more care should be used in cleaning around it.

Preparing the Patch. Having the hole well cleaned and ready, these cleaned parts should be painted with two coats of vulcanizing cement, which is allowed to dry. This must be thoroughly, not partly, dry. Then the proper patch is selected, the smaller size being sufficient for small patches, while in the case of large repairs, the patch should be from $\frac{1}{2}$ to 1 inch larger all around than the hole. If this is not a prepared patch, one side should be cemented just as the tube was previously. If a prepared patch is used, the semi-cured side should be placed in, that is, with the sticky or uncured side toward the tube from the inside.

When the cement on the patch is just sticky enough, it should be inserted and the tube pressed down against it all around, slowly and carefully so as to get good adhesion. Next the cavity about the inside patch is filled with gum or pure rubber, preferably in sheet form as it comes for this purpose. This is filled in until the surface is flush. It is preferable to use a little vulcanizing cement to hold this rubber in place, particularly if a piece of sheet gum is cut to fill the hole.

Vulcanizing the Patch. The repair is now about half completed and is next vulcanized. The length of time, if steam is used, varies with the amount of steam pressure; if the portable gasoline or alcohol type of vulcanizer is applied the time varies with the temperature. As this time variation is so wide, it is impossible to give an invariable rule. Thick tubes require a little longer than thin ones, large patches longer than small ones, wide patches more than narrow, etc. The vulcanizing must be carefully and thoroughly done, and, as the success of the whole job depends upon this one process, the arrangement of the tube on the plate, of the soapstone on the new rubber and on the vulcanizer to prevent adhesion, of the wood or rubber pad above the patch, of the clamp and its pressure, should all have careful attention. With 60 pounds steam pressure available, from 10 to 12 minutes is about right, with 75 pounds from 8 to 10 minutes. In any case, the rubber should be cured just firm enough not to show a slight indentation from the point of a lead pencil. This is a good test to use at first, although after a short experience, the workman will be able to judge of the condition from the feeling, color, and general appearance of the patch.

When the size of the plate is small, the tubes should be held up above it out of the way, partly to allow the full use of the plate surface, but also to keep the tubes from being damaged.

Inserting New Section. *Preparing the Tubes.* In case the damage to the tube is too great to permit the use of a patch, for instance, in case a blowout makes a wide hole perhaps 7 inches or more long, in an otherwise good tube, it is advisable to cut out the damaged section and insert a new section in its place. Sometimes old tubes of the same size can be used for this, but, if not, sections can be purchased from the larger tire and rubber companies.

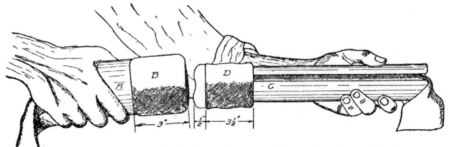

Fig. 29. Sketch Showing Method of Inserting New Section in Inside Tube

In the repair, proceed as follows: After cutting out the damaged section, bevel down the ends very carefully, using a mandrel to work on and a very sharp knife. As the appearance and, to a large extent, the value of the repair will depend upon these beveled ends, this should be done in a painstaking manner. Next select the tube section and cut it to size, that is, from 5 to 6 inches longer than the section which was cut out and which this patch is replacing. This allows $2\frac{1}{2}$ to 3 inches for the splice at each end. Bevel the ends of the tube as well, and, after beveling all four ends, roughen them with a wire brush or sandpaper.

Making the Splice. Having the tube and repair section beveled and buffed, the ends to be joined should be coated with one heavy or two light coats of acid-cure splicing cement. With the tube and patch properly placed on the mandrels—tube on the male and patch on the female—turn back the end to be repaired and the end to be applied as shown in Fig. 29. At *A* is shown the female mandrel on which is the patch *B*, turned back from the end of the mandrel about the right distance, say 3 to 3½ inches. On the male mandrel *C* the tube *D* has been turned back about 7 to 7½ inches, then turned back again on itself about 3 to 3½ inches.

Just as soon as the cement has dried thoroughly on the tube, apply a coat of acid to the patch and immediately place the two mandrel ends together and snap, blow, or push the end of the patch over on to the end of the tube. This frees the female mandrel, which can be laid aside. Immediately wind the patched portion (still on the male mandrel) with strips of muslin or inner tubing. In 15 to 20 minutes the cement will have formed a permanent union, the wrappings can be removed, and the tube withdrawn through the slot in the mandrel.

This done successfully, the whole operation is repeated for the other splice. If the splice does not cure together well, it indicates either that the acid supply is poor or else the splicing was not done quickly enough after applying the acid.

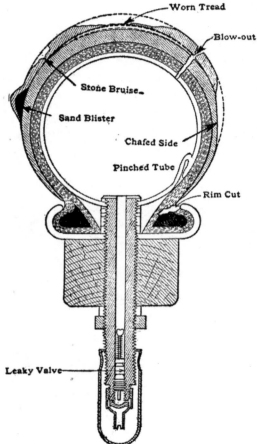

Fig. 30.   Section of Tire, Showing Forms of Troubles

Outer Shoe, or Casing, Repairs

Classifying Troubles. Some of the common tire troubles— those of the inner-tube variety just discussed, and casing troubles as well—can be clearly shown by suitable illustrations. For example, a section through the tire showing how the troubles occur is sometimes very useful, as shown in Fig. 30. Here the pinched tube and blowout are indicated, the results of these on the inner tube and also their method of repair having just been described. These troubles together with punctures, leaky valves, and porous rubber in the tubes about cover the extent of inner tube troubles. Because of their more complex construction, casings have more numerous and more varied troubles, which, consequently, are more difficult to repair. The more common casing troubles are blisters, blowouts, rim cuts, and worn tread, the latter indicating the necessity for retreading. These will be described in order.

Sand Blisters. The sand blister shown on the side of the tire, Fig. 30, is brought about by a small hole, such as an unfilled puncture hole, in combination with a portion of the tread coming loose on the casing near this hole. Particles of sand, road dust, dirt, etc., enter, or are forced into, this hole and move along the opening provided by the loose tread. Soon this becomes continuous and the amount of dirt within the break forces the surface rubber out in the form of a round knob known as a sand blister. This is cured by cutting open the blister with a sharp knife on the side toward the rim and picking out all dirt within. When the recess is thoroughly cleaned, the hole and the radial hole in the tire tread nearby should be filled with some form of

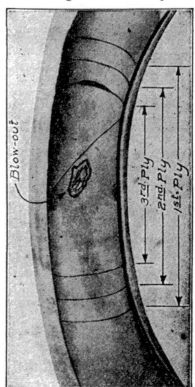

Fig. 31. Method of Preparing Layers of Fabric for Patching Blowouts— Inside Method

self-curing rubber filler, a number of kinds of which are sold. The double benefit of this is to close the hole so that the trouble is not repeated and to keep out moisture which would ultimately loosen the entire tread.

Blowouts. The blowout, which is perhaps the most important casing repair, may be made in two ways: the inside method, in which the whole repair is effected on the inside; the combination inside and outside method.

Inside Repair Methods. Refer back to Fig. 30 for the general tire construction and to Fig. 31 for this particular case, the inside of the tire is held open by means of tire hooks and the inside fabric layers or plies removed for a liberal distance on each side of the opening. As shown in Fig. 31, a lesser amount of the second layer should be taken than of the first, and still less of the third and each subsequent one. On 3½- and 4-inch tires it is not advisable to remove more than two plies; on 4½-inch tires three, as shown; and on the larger sizes four plies. The edge of each layer of fabric should be beveled down thin, as well as the material directly around the blowout.

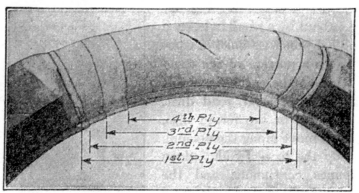

Fig. 32. Method of Preparing Fabric for Blow-Out Patch—Inside and Outside Method

Apply a coat of vulcanizing cement and when it has dried, say for an hour, apply another. When this has dried enough to be sticky or tacky, fill as much of the hole as possible with gum. When this is filled in level, apply the fabric patch. This is made up to match the fabric cutout, that is, if three layers are removed, it should consist of three plies stepped-up to match, and an extra last ply of bareback fabric unfrictioned on one side. This last layer should extend 3½ to 4 inches beyond the ends of the patch.

When this is properly applied and carefully smoothed down, the tire is placed in a sectional mold, clamped in place, perhaps wrapped with muslin strips to hold it tightly against the mold, and heat applied from the inside. This makes an excellent repair and a fairly quick and easy one, but it is not applicable for large blowouts; at least, it is not as effective as the inside and outside method.

Inside and Outside Method. In the inside and outside method, the material is removed from the outside, stepped down, and beveled in the same manner as for the method just described. Fig. 32 shows a tire with a medium size blowout, which has been stepped down for a sectional repair, four plies having been removed. The rule for the number of plies to remove is about the same as before, except that in the larger sizes this should depend more on the nature of the injury. It should be noted, however, that in this case the plies have all been removed right down to and including the bead. This is done to give the new fabric a better hold and to make a neater job and one that will fit the rim better. Give the whole surface two good coats of vulcanizing cement, allowing it to dry thoroughly.

Apply the same number of plies of building fabric as were removed, with the addition of chafing strips of light-weight fabric at the bead. Over this building fabric apply a thin sheet of cushion gum, slightly wider than the fabric breaker strip; then a thickness of fabric breaker strip over this; and then over this fabric another sheet of gum, slightly narrower than the previous sheet. All this, however, should be built up separately and applied as a unit and not one at a time, as described. These several plies should be well rolled together on the table. All edges should be carefully beveled off, especially the edges of the new gum where it meets the old, as it is likely to flow a little and leave a thin overlap which will soon pick loose.

No fabric is removed from the inside, but the hole is cleaned, its edges beveled, then filled with tread gum, and the inside reinforced with a small patch of building fabric; over this lay two plies of building fabric of considerable size. Now the whole casing is placed in a sectional mold, a surface plate applied to the outside, and heat applied both inside and outside. This will heat the tire clear through and make a good thorough job of curing.

Rim=Cut Repair. *Partial Cut.* To repair a partial rim cut, one or two plies of the old fabric are removed, unless it is severe, when three plies may be taken off. This is removed right down clean as explained under Blowout Repairs, and the cement and new materials applied in the same way, with the omission of the fabric breaker strip. However, care should be used to carry all building fabric layers not only down around the bead to the toe but up on the inside far enough to secure a good hold and ample reinforcement. If this should make the rim portion somewhat more bulky, remember it was a case of doing this or getting a new tire.

Complete Rim Cut. Where the rim cutting is continuous, the old side-wall rubber is removed up to the edges of the tread, and the old chafing strips and one ply of old fabric to about an inch above the beads removed also.

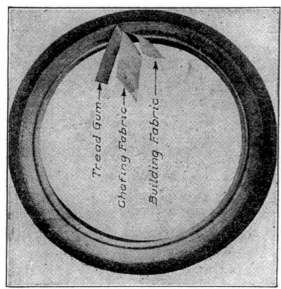

Fig. 33.   Method of Handling Rim Cuts

Cut through the side-wall rubber all around, but be very careful not to cut into the fabric body, or carcass. The whole of the side wall and chafing strips can be removed in one operation. Apply two coats of cement and, after this is thoroughly dry, put on a patch consisting of one ply of building fabric, one ply of chafing strip, and a surface, or outside, ply of new tread gum. This is made on the table and the parts thoroughly rolled together. When completed, vulcanize in a sectional mold with sectional air bag and bead molds or endless air bag; apply to a split curing-rim wrap, and vulcanize in heater or kettle. The tire is repaired, but not vulcanized, and, with the ends of the three applied plies of material loosened to show, may be seen in Fig. 33.

Retreading. Retreading is a job which must be done very carefully, not only because of the job itself, but also because this is probably the most expensive single job which can be done to a tire, and the worker should make sure before starting that the wire warrants this expense. It should have good side walls and bead, and the fabric should be solid and not broken apart.

Repairing the Carcass. In the usual case, it is advisable to remove not only the surface rubber and fabric breaker strip, but also the cushion rubber beneath the breaker strip, that is, the tire should be cleaned off right down to the carcass, and the latter cleaned thoroughly. As the rubber sticks, a rotary wire brush will be found useful and quick. However, this should be used carefully so as not to gouge the carcass. After buffing, the loose particles of rubber should be removed with a whisk broom or dry piece of muslin. In this cleaning work the carcass should be kept clean and dry. Apply two coats of vulcanizing cement and allow both to dry; the first should be a light coat to soak into the surface fabric; the second should be a heavy coat.

Building Up the Tread. In building up the tread, it should not be made as heavy as the former tread, as the old worn and weakened carcass cannot carry as heavy a tread as when new. Furthermore, it takes longer to vulcanize a heavy tread and presents more opportunity for failure. In the building-up process, the proportioning of weights is important, and should be taken from the tabulation below, which represents years of experience in tire repairing:

Size of Case (in.)	Ply toward Fabric (in.)	Second Ply (in.)	Third Ply (in.)	Fourth Ply (in.)	Fifth Ply (in.)	Last Ply Over All	Complete Tread Consists of
3	$2\frac{3}{4}$	$3\frac{1}{2}$				*See Note	3 plies
$3\frac{1}{2}$	$2\frac{3}{4}$	$3\frac{1}{2}$	$4\frac{1}{4}$			*See Note	4 plies
4	$3\frac{1}{4}$	4	$4\frac{3}{4}$			*See Note	4 plies
$4\frac{1}{2}$	4	$4\frac{3}{4}$	$5\frac{1}{2}$			*See Note	4 plies
5	$4\frac{1}{4}$	5	$5\frac{3}{4}$	$6\frac{1}{2}$		*See Note	5 plies
$5\frac{1}{2}$	$4\frac{3}{4}$	$5\frac{1}{2}$	$6\frac{1}{4}$	7	$7\frac{3}{4}$	*See Note	6 plies
6	$5\frac{1}{4}$	6	$6\frac{3}{4}$	$7\frac{1}{2}$	$8\frac{1}{4}$	*See Note	6 plies

* Note—Determined by condition of case after buffing and cementing.

Size of Case (in.)	Width of Breaker Strip (in.)
3	$1\frac{5}{8}$
$3\frac{1}{2}$	$2\frac{1}{8}$
4	$2\frac{1}{2}$
$4\frac{1}{2}$	3
5	$3\frac{5}{8}$
$5\frac{1}{2}$	4
6	$4\frac{1}{2}$

This tread strip is built up on the table with exceeding care, all edges being rolled down carefully. When the strip has been prepared and the carcass is ready for it, one end should be centered on the carcass, and then the balance of the strip applied around the circumference, being careful to center it all around, as the workman in Fig. 34 is doing. After it has been applied all around, it should be rolled down carefully, all air pockets opened with a sharp pointed awl, and the gum at the edges of the plies rolled down with the corrugated stitcher. When ready, vulcanize in a kettle, using an endless air bag with tire applied to a split curing-rim, and wrapped—preferably double wrapped—all around.

Use of Reliner. Many a casing which appears good on the outside but which really is unsafe because of fabric breaks on the inside can be saved, or its life temporarily prolonged, by the application of a reliner. By this is not meant the prepared canvas and fabric reliner which can be put in dry, but a regular built-up strip of building fabric, vulcanized in place so as to be an integral part of the tire. For ordinary breaks, use a single ply of building fabric on a casing which has been entirely cleaned out and which has had two coats of vulcanizing cement thoroughly dried

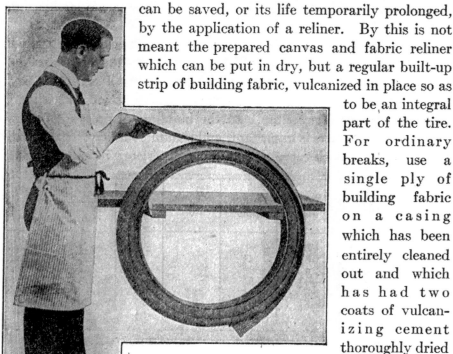

Fig. 34. New method of Putting on New Tread

in. In the case of a bad break, use two plies of fabric, stepping them to fit; the under ply should be frictioned on two sides and coated on one, and the upper ply should be frictioned on one side only, the side toward the tube being bareback. Use an endless air bag for internal pressure, apply to a split rim, wrap, and vulcanize in a kettle from 35 to 45 minutes at a steam pressure of 40 pounds.

Summary. By the application of parts of the foregoing instructions and the use of much common sense, coupled with a knowledge of the construction, use, and abuses of tires, the repair man will be able to handle any form of tire repair brought to him. In starting out, perhaps he could not do a better thing than to take an old tire apart to see just how it is constructed. This will give a much more clear idea than any number of diagrams, sketches, or photographs.

The tire repair man should remember, too, that this is no longer a game, but that, by means of scientific apparatus and the application of correct principles, it has been brought up to a high state of perfection; an expert can predict with reasonable accuracy what will happen in such and such a case, if this and that are not done. In short, the tire-repairing business within the last few years has been brought up to a stage where it, or any part of it, is a dependable operation. The tire repair man should handle all his work from this advanced point of view; it will pay the largest dividends in the long run.

REBORING EQUIPMENT
Courtesy of Gisholt Machine Company

RIMS

Kinds of Rims. Nearly all rims are of steel or iron, but vary greatly as to types. Naturally, the first rims were of the plain type, while the latest are of the demountable, remountable, or removable types, all these being very much the same. Between the two came the clincher rim, which is properly a plain rim; and the quick-detachable rim.

Plain Rims. The form of rim first used was naturally the solid type. This form is a simple endless band with two edges just high enough to prevent the tire from coming off sidewise when it has once been stretched in place. Nothing like it is used today, the nearest approach being the form of rim used with single-tube bicycle tires.

Clincher Rims. Clincher rims were brought out primarily to avoid the weaknesses of the Dunlop, viz., a weakness at the base, and, hence, it had an unusually heavy bead. Another fault which this tire remedied was the tendency under high pressure for the tire to draw away from the rim. This was avoided by the edge of the clincher being made fairly wide where it was designed to go into the pocket, or groove, formed by the contour of the rim.

It is the depth of this pocket, or groove, and the corresponding size of the edge of the bead on the tire, both excellent qualities, which make the tire hard to put on and take off.

Quick=Detachable Tire Rims. It was this inherent difficulty of handling the clincher tire and rim which brought about the quick-detachable tire. This did not differ from the clincher tire in the tire portion, the difference being in the rim, which has one curved portion made in removable form, with a locking ring outside of it or made integral with it. In some quick detachables, the rim is expanded by a special tool and a spacing piece set into place, which holds the edge expanded. When this is done, the ring—as it is a simple ring with special ends—is held in place until released by the use of the special tool. On the end of the ring there are two little square lugs which project downward and have a hook shape. The one edge of the rim, made flat and straight on that side, has a slot with staggered, rectangular ends into which these lugs fit. It requires force to spring the rings together so the lugs will go into the slots, but once in place, the natural springiness of the rings holds them firmly in place, and holds the tire as well.

Figs. 1, 2, and 3 are given to show how this ring is put in place on a tire. Fig. 1 shows the beginning of the operation, and the instructions for the different steps will make them clear. Thus:

Always start with left end of the ring. Lock this in the rim as shown in Fig. 1, so that the end of the ring is flush with the slot provided for the second end. A dowel pin is provided to register the ring in the proper place. This must always be correctly centered or the ring cannot be applied. This done, the balance of the ring can be forced over the flange of the rim, as shown in Fig. 2, with the exception of the locking end. By means of the tool, the last locking end can be

Fig. 1.   Putting on a Q.D. Ring.
The Start

Fig. 2.   Putting on a Q.D. Ring.
Forcing Flange over Rim

raised and forced over the rim into the recess provided for holding the same in position preparatory to drawing the ends together, Fig. 3, showing the correct position of the tool.

Then by entering the two points of the tool in the holes provided in the ring, the ends may be drawn together, as shown in Fig. 3, and, with a slight additional leverage, the ends of the rings can be made flush.

Before proceeding further, it should be stated that the object of the quick-detachable rim is the quick removal of the tire, in order to allow a quick repair or substitution of the inner tube. On the other hand, the object of the demountable, remountable, removable, and other rims is the removal with the tire of the rim itself to allow the substitution of a new tire and rim, the tire being already inflated and ready for use as soon as applied. The object of the removable

wheel is the removal of the entire wheel with rim and tire in order to substitute a spare wheel with already inflated tire.

It might be thought that these methods called for the carrying of extra weight, but the amount added is actually very small, as, by their use, tire tools and pump are dispensed with and their weight saved.

Fig. 3. Putting on a Q.D. Tire.
The Locking Ring

Fig. 4 shows the former Goodyear rim. This rim, as will be noted, is of the quick-detachable type, the idea being to remove the tire only. The rim itself has a button-hook shape with a slight ridge, or projection, answering to the handle. This is on the fixed side, the inner flange inside of the tire butting against it as a stop. The tire is pushed over against this, being held

on the outside by a second flange of similar shape. The latter, in turn, is fixed in place by a locking ring, a simple split circular ring of deep oval section. This fits into the button-hook portion, its contour being such as to fit it exactly. In use, it is sprung into place, the outer edge of the hook on the rim and the natural spring of the ring preventing it from coming out. This makes a very simple and serviceable quick-detachable rim. To make doubly certain that the locking ring cannot jump out, a

Fig. 4. Former Goodyear Universal Rim

spreader plate is attached to the valve stem; screwing this down into place wedges the bead of the tire over against the outer flange, which, in turn, pushes the locking ring tight against the outer curved part of the hooked rim. When in this locked position, the upper part of the flange hangs over the locking ring, so that it cannot rise vertically, the only manner in which it could come off. This rim is shown with a detachable tire in

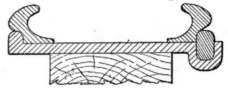

Fig. 5. Adapting Goodyear Rim to Clincher Tires

position, but may be used with any standard clincher tire by the use of extra clincher flanges. Fig. 5 shows the rim with a set of these flanges in position, ready to take a standard clincher tire.

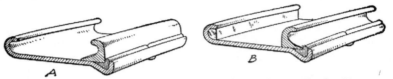

Fig. 6. Universal Q.D. Rim No. 2 Arranged for Clincher and Dunlop Tires

Quick-Detachable Number 2. Figs. 6 and 7 show the standard quick-detachable rim, now known as No. 2. This was adopted by the Association of Licensed Automobile Manufacturers

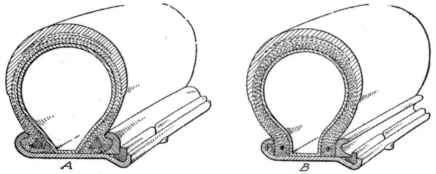

Fig. 7. Universal Q.D. Rim with Tires in Place

as a standard and given the above name. It has the feature of accommodating all regular clincher, or Dunlop tires. In Fig. 6, it is shown at *A* ready for a clincher tire and at *B* ready for a Dunlop tire, the adaptation for the straight sides being shown.

The two parts of Fig. 7 show sections of tires in place, making clear the exact use of this reversible flange. *A* shows a regular clincher tire in place, while *B* reveals the reversed flange in place with a Dunlop tire. Both Figs. 6 and 7 show the construction of

320

the device, the outer dropped portion of the rim having a hole through it. The locking ring is split vertically and one end, just at the split, carries a projection or dowel pin extending downward. To put the rim on, this dowel pin must be fitted into the hole in the rim to give a starting place. When this has been done, one may force the balance of the ring into place around the wheel with any suitable, thin, wedge-shaped tool.

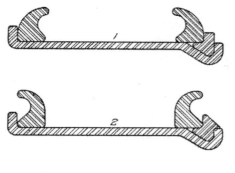

The shape of this locking ring with a right-angled groove in its inner edge permits the outer flange to overlap it, which insures the retention of the ring when once it has been put in place. Furthermore, it gives the outer side flange a wider seat on the rim, thus making it more stable and longer wearing.

Fig. 8. Sections through Three Popular Q.D. Universal Rims

As will be noted, the difference between these two rims—that is, the old Goodyear and the Universal No. 2—lies in the saving of one ring and the shape of the locking ring. Both of these are called universal rims because they may be used interchangeably for straight-

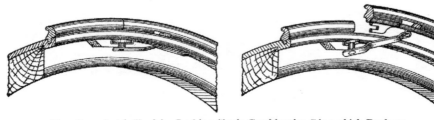

Fig. 9. Latch Used for Locking Single Combination Ring which Replaces Former Side Ring and Locking Ring

side and clincher types of tire. Other Q. D. Universals are shown in Fig. 8, although, in the opinion of tire men, the Universal form is slowly going out of use.

To explain these briefly, No. 1 is a modification of the Goodyear, with different shaped inner rings, while the locking ring and the lip formed in the felloe band to receive it are similar to those of Universal No. 2. In *2* the only difference from *1* lies in the locking ring, which has a modified Z-section, with a lip extending over the outer edge of the felloe band. The third section differs from the other two only in having the outer ring and locking ring combined into one, and the felloe band changed to suit this. This combination ring is held in place by means of a simple swinging latch, which is shown open and closed in Fig. 9. When opened, this permits raising the end of the ring, to which the shape of the felloe band offers no resistance. The whole inner ring is taken off, following around the circumference of the wheel, after which the tire is easily removed.

Quick-Detachable Clincher Forms. To return to the plain clincher tire and the Q. D. rim, which allows of its ready removal,

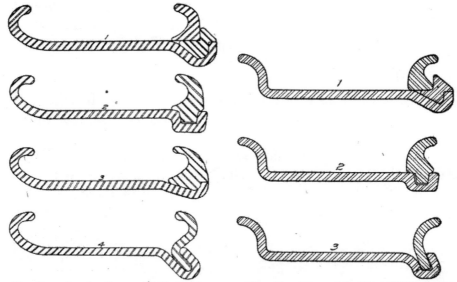

Fig. 10.   Popular Forms of Q.D. Clincher
Rims, Shown in Sections

Fig. 11.   Three of the Most Widely Used
Straight Side Q.D. Rims

Fig. 10 shows four of the most prominent forms, these being indicated simply as flat sections of the rim, for the tire is the same in all cases. All these have the simple clincher edge on one side, with removable ring and locking device on the other. That at *1* has the same locking device shown at *2* in Fig. 8, the Z-shaped ring extending over the edge of the band. That at *2* is practically the same as *3* in Fig. 8. The one seen at *3* is similar to that at *2* except for the detailed shape of the ring as well as the lock (not shown). The advantage of the form shown at *4* is that the outer ring is self-locking, that is, the shape of ring and band are such that when the former is in place the tire itself locks it. Its only disadvantage is that it is harder to operate than the other forms, yet despite this fact it has been recommended for general adoption as the only Q.D. clincher rim worth continuing.

Q.D. Type for Straight Sides. To close the subject of straight side tires, the rims of the quick-detachable form now in use aside from those already shown are seen in Fig. 11. Here these are seen to be identical with *1, 2,* and *4* of Fig. 10, except that the fixed side is arranged for a straight side instead of being made with a clinch. Here again, the last form of self-locking type has been recommended as a standard.

Demountable Rims. All, or practically all, demountable rims come under one of two headings—those in which the tire can be detached on the wheel without demounting (if it is so desired) and

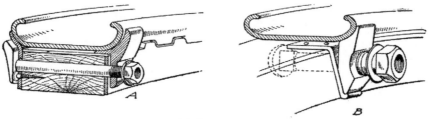

Fig. 12. Sections of Michelin and Empire Demountable Rims

those which are of the transversely split type and must be demounted before the tire can be removed. In addition, there is a second division of demountable rims into those which have a local-wedge form of attachment and those which have a continuous holding ring, this, in turn, being held by means of local wedges. Any of the plain demountables, which will be called demountables from now on, may be of either type of attachment, as is also the case with the first-named or demountable detachables.

Local Wedge Type. In the so-called local wedge type, which includes the well-known Continental forms (notably Standard Universal Demountable No. 3 and Stanweld No. 22 and No. 30), Michelin, Empire, Baker, Detroit, Prudden, Standard Universal Demountables No. 1 (formerly the Marsh), and No. 2, and others, loosening the six (or eight, as the case may be) bolts frees the rim directly without further work. In some of these, such as the Michelin; the various Continentals, including Stanweld No. 22 and No. 30;

Detroit; Baker; and others, the wedges carry a projecting lip, which makes it necessary to unscrew the nuts far enough to allow the

removal of the wedge so as to pick this lip out from under the tire-carrying rim. In others, such as Empire, S.U. No. 1 and No. 2, the construction of the wedge and rim is such that loosening them frees the rim, the upper part of the wedge or clip swinging down to the bottom position as soon as loosened, because of its heavier weight and the fact that there is no projecting edge to prevent it. While this latter construction makes a faster operating rim, it is an open question as to whether it is as safe as the other form. These two constructions are shown very plainly in Fig. 12, in which *A* is the Michelin with lipped wedges, and *B* the Empire with plain wedges.

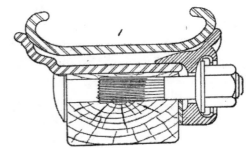

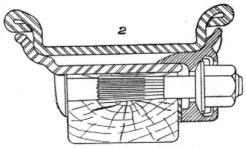

Fig. 13.   Two Popular Demountable Rim Forms —for Clincher Tires above, for Straight Side below

In Fig. 13 is shown a pair of additional demountables, which are held by the local wedge method, the difference here being in the form of a wedge. Note that *1* has a solid clincher rim and *2* a straight side rim. The base, however, is the same for both and, as will be seen by examining this, has two

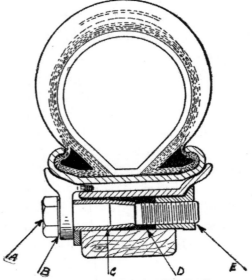

Fig. 14.   Sectional Drawing Showing Construction of Baker Demountable Rim

curves in its upper surface, the straight side rim fitting into the lower or bottom one, while the clincher form of rim fits into the upper one. Note, also, that the wedges are the same for these two. This makes the demountable parts of the rim practically universal in that the owner can change from clincher to straight side or *vice versa* by simply purchasing the extra set of tire-carrying rims, no change in the wheels or means of attachment being necessary. For this reason, the felloe band shown under these two rims has been suggested as a standard for demountables.

Fig. 15. The First Operation in Removing Baker Demountable—Loosening the Bolts

Process of Changing Baker Local Wedge Type. In Fig. 14 is shown the Baker, which, as mentioned previously, is of the local wedge type of demountable, having a transversely split rim which must be removed from the wheel before the tire can be taken off. Perhaps this whole action will be shown more clearly by the progressive series of views, Figs. 15 to 25, which show the various steps in removing and replacing a tire and tube mounted on a Baker rim, the same as is shown in section in Fig. 13.

Fig. 16. Second Baker Demounting Operation— Jacking the Wheel and Starting to Pry off Rim

First, all the wedge bolts except the two nearest the valve stem, one on either side, are loosened by means of the special brace until the wedges swing out and down, as shown in Fig. 15. As mentioned previously, this means quite a little loosening, for the

wedges have a long lip which projects under the tire-carrying rim. When this has been done, and as each one swings down out of the way, it is tightened just enough to prevent the wedges from swinging back.

This done, the wheel is jacked up off the ground, as shown in Fig. 16, and the point of the tire tool is inserted between the felloe band and the rim carrying the

Fig. 17   Third Baker Operation—Putting on New Tire and Lowering Wheel

tire at the point opposite the valve, where, it will be remembered, the wedges were loosened, and the rim will be almost free. By prying

the tire-carrying rim outward and working around it toward the valve and back again, it will finally be loosened to a point where, with the valve at the bottom, the rim and tire can be slipped off without lifting it. The extra tire and rim are now put in place.

This is shown in Fig. 17, where the reverse of the operations shown in Fig. 16 and just described is followed, that

Fig. 18.   Fourth Operation—Tightening Bolts on the New Tire and Rim

is, the valve stem hole is revolved to the top, the valve stem inserted, the rim pressed into place all around, then the wheel is revolved until

Fig. 19.   Fifth Operation—Starting to Take the Rim out of the Tire—Beginning to Pry Short End

Fig. 20.   Sixth Operation—Forcing Down the Short End of Rim

Fig. 21.   Seventh Operation—Prying under the Loose End of Rim

Fig. 22.   Eighth Operation—Raising the Free End of Rim, Using Both Hands

Fig. 23.   Ninth Operation—Inserting Valve Stem and Beads in End of Rim

Fig. 24.   Tenth Operation—Prying Tire Away from Rim to Let Latter Slip into Place

the valve stem comes to the bottom, so that the two wedges which have not been loosened are nearest the ground. Then the jack is let down and removed, the whole weight of the wheel coming on the bottom point where the wedges are already tight, never having been loosened.

This action is necessary as, with the weight on the other points where wedges are still loose, it would be necessary to work against the car weight. At this point, as Fig. 18 shows, the nuts are loosened, using the special brace until the wedges can be inserted under the rim. This done, the nuts are tightened to hold them there. This tightening is continued until the little studs, or lips, in the rim rest on top of the outside edge of the felloe band, using the tire tool to force them in, if necessary. The new tire carried is supposed to be ready for use, that is, inflated to the proper pressure, so that these four actions complete the work of making a roadside change.

When it is desired to repair the tire which has been removed, it is carried home on its rim just as taken off the car wheel, and the rim is removed from the casing as follows: Rim and tire are laid flat on the garage floor, as shown in Fig. 19, so that the outer end of the diagonal cut in the inside of the rim which is farthest from the valve stem is uppermost. An inside plate will be found on the rim which covers the two rivet heads on either side of the cut, with a central hole for the valve stem. This plate is called the anchor plate and must be removed. To do this, begin at the short end of the rim, which does not have the valve stem—as, in this position, it will be held in the long end—and insert the sharp end of the tire tool or a screwdriver under the bead or between the bead and the rim.

These two actions, as shown in Fig. 20, bring the two short sides of the rim closer together and thus reduce the diameter. When the extreme end has been freed in this way, the operation is repeated some 5 or 6 inches farther around, that is, that much farther away from the slit. This done, a considerable portion of one end will be free. Then turn the rim and tire over so that this free part comes at the top instead of at the bottom and, standing on the part which is still tight, insert the tool between the rim and the entire tire.

This frees the entire end, but, to make sure, the tool must be moved a little farther along so as to free more of it. When enough has been freed to allow grasping it with both hands, as shown in Fig. 22, the tool is dispensed with and, taking a firm grip on the rim, at the same time standing on the tire at the point where tire and rim still contact, pull upward strongly. When followed all the way around this pulls the rim entirely out of the tire.

Having the casing and tube free, they may now be inspected and repaired. When this is done, or if it is not done, and a new tire or tube or both are used, the worker is ready now to replace the rim. This is practically the reverse of the method just followed out. As shown in Fig. 23, the rim is laid on the floor; then the end which has the valve-stem hole drilled in it is raised, and the valve stem inserted. Next the beads are pulled into the rim, it being necessary to press them together somewhat tightly in order to do this, but, with a little practice, it soon becomes an easy matter. All this is done with the other part of the rim underneath the tire.

The inserted end of the rim is followed around with the thin end of the tire tool, as shown in Fig. 24, the position of the tire

Fig. 25. Eleventh Operation—Inserting Anchor Plate

above the rim allowing the workman to stand on it and thus use his weight to press the two sides of the tire together and, at the same time, to force them into the rim. This operation is followed right around the inside circumference of the tire, the free, or short, end of the rim being the last part to enter. On account of the shape of the joint or cut in it, this should slip readily into its proper place, but if it does not, the thin end of the tool can be used to pry it into place, or a hammer can be used on the longer side to drive it in.

The rim being fitted snugly into place all around, the anchor plate is inserted, Fig. 25, to prevent the short end slipping out again, and the tire is ready for inflation. If it is to be carried as a spare tire, the dust cap should be screwed into place over the valve stem, so as to preserve the threads which might be damaged in handling.

Rim with Straight Split. This covers the action of practically all the demountables in which the transversely split rim is used, necessitating the removal of the rim and tire from the wheel before the tire can be taken off the rim. However, not all rims are split on a diagonal as is this one, and Fig. 26 is presented to show this single feature on another rim, which otherwise is somewhat similar. Here

the rim is split at right angles, having a plain thin rectangular plate *A* attached to the free end, or that which is removed first, while the other end has a swinging flat tapered plate with a cam-shaped end *B*, the action of which is to expand the rim to its fullest diameter and lock it there. In the top figure, it is locked—that is, the rim is expanded as it would be when in use and just after it had been removed for replacement. When the rim is to be removed from the tire, the latch *B* is swung out of the way, as shown in the lower figure, when the catch *C* which holds the two ends together can be opened by lifting the tire with this portion at the bottom and then dropping it a couple of times. This done—usually this action will be accompanied by the free end spring inside the fixed end—continuation of the removal is an easy matter. The rim shown is the Stanweld No. 20.

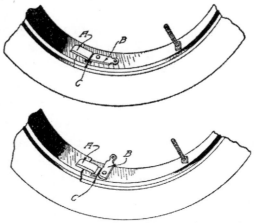

Fig. 26.   One-Piece Rim, Showing Right-Angled Split and Locking Device

Courtesy of Standard Welding Company, Cleveland, Ohio

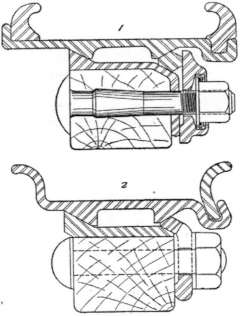

Fig. 27.   Sections through Two Popular Forms of Demountable-Detachable Rims

Comparison of Continuous Holding Ring Type with Local Wedge Type.   To return to demountable-detachable rims, these may and do include a number of those quick-detachable forms previously shown and described. In Fig. 27, a pair of typical forms is shown, that at *1* being fitted for a clincher tire, while that at *2* is for a straight side. Looking at the detachable part of the rim, *1* will be recognized as that previously shown at *3*, Fig. 8, where it was described as a universal rim, the inversion of the two rings converting it from a clincher to a straight side, or *vice versa*. Similarly, *2* will be recognized as the form of detachable shown at *3* in Fig. 11.

Here, however, both are fitted to be used as demountables, this being accomplished by the formation on the under side of the band of a pair of wedge-shaped projections. The felloe band is so made and applied that it forms one surface to contact with one of these wedges, while the other is formed variously. At *1*, a separate ring is used with the flat outside clips to hold this against both felloe band and rim, while at *2* the wedges or clips have an extension which presses against the outer wedge on the rim. This latter distinction divides these two into the two classes mentioned previously — one into the continuous holding ring class, the other into the local wedge type.

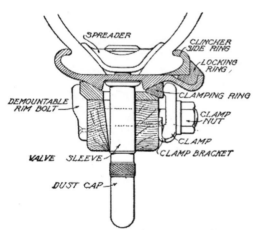

Fig. 28. Section of Tire and Rim of Firestone Demountable Tire

These forms are shown to illustrate this point and also because, despite this difference, they have practically similar felloe bands. This felloe band—that is, of the form shown in *2*—has been recommended as a standard for all demountable-detachable rims. Another and different example of the clamping-ring demountable-detachable type is shown in Fig. 28, this being the Firestone rim. Here, it will be noted, is the felloe band just mentioned, while the detachable-rim portion is that previously shown at *1* in Fig. 10 as having the Z-shaped locking ring and being adapted to clincher tires only. The rim band is made with the two wedge-shaped projections on its underside.

Perlman Rim Patents. Late in the summer of 1915, considerable consternation was caused among tire and rim manufacturers when it became known that the Perlman rim patent had been adjudged basic by the courts, and that, on the strength of this decision, an injunction had been issued against the Standard Welding Company, of Cleveland, Ohio, some few of whose rims have been previously described. Perlman's original patent was applied for on June 29, 1906, and, in addition to this record, the fact was established that the owner had a Welch car which had traveled over 150,000 miles and on which were a set of the original rims. The case dragged through the courts and was discontinued some seven or eight years ago. Perlman persisted, however, although he had to revise and alter his application many times; the basic patents were finally allowed, and issued to him in February, 1913. This means, of course, that the patent will not expire until the year 1930.

Perlman's locking elements and the principle involved are shown in Fig. 29, which is a section through the rim and felloe. In Perlman's suit, it was claimed that the wedge end of the bolt which was covered in his patent, included all wedge-operating rims, whether actuated from the center, as in Fig. 29, or from the side. This contention was supported by the

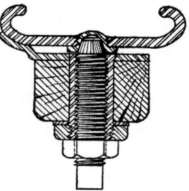

Fig. 29. Section of Perlman Rim, Showing Locking Device

court, and negotiations are now in process between Perlman and many manufacturers of the so-called local wedge type of rim. As this would appear to cover all the rims shown and described in Figs. 12 to 28, inclusive, the influence of this decision upon the industry can be imagined. Moreover, the length of time which this basic patent has to run precludes the possibility of delaying action by prolongation of suits, as has been done in similar cases. A notable example of this is the case of the Selden automobile patents, which were fought on one ground or another over a long period of years.

Standard Sizes of Tires and Rims. As might have been noted in going over the above discussion of tires, plain rims, detachable rims, and, finally, demountable rims, all these different constructions require widely differing wheel sizes. It has been proposed to standardize wheels, that is, the outside diameter of the felloe and with it the thickness of felloe bands as well as their shapes or contours, one for each tire cross-section. The proposed reduction of tire sizes to nine standards is as follows: 30- by 3-inch, 30- by 3½-inch, 32- by 3½-inch, 32- by 4-inch, 34- by 4½-inch, 36- by 4½-inch, 38- by 5½-inch and probably 36- by 5-inch, supplying these sizes and these only to manufacturers of cars; additional oversizes are allowed for car users, one for each size above, that is, 31- by 3½-inch for 30- by 3-inch, 31-

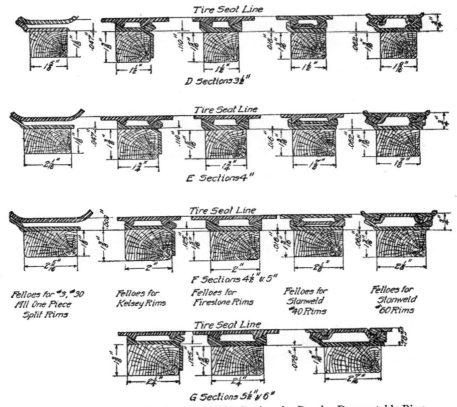

Fig. 30. Typical Felloe, Band, and Rim Sections for Popular Demountable Rims

by 4-inch for 30- by 3½-inch, 33- by 4-inch for 32- by 3½-inch, 33- by 4½-inch for 32- by 4-inch, 35- by 4½-inch for 34- by 4-inch, 35- by 5-inch for 34- by 4½-inch, 37- by 5-inch for 36- by 4½-inch, 39- by 6-inch for 38- by 5½-inch and probably 37- by 5½-inch for the 36- by 5-inch. Rim standardization will follow the adoption of these sizes. In this event, the standardization of demountable rims will come in time.

At the present, there is a wide range of difference, as will be noted in the drawing, Fig. 30, which shows felloes for the most widely used demountable rims, depicting the band and rim in each case. The drawing should be read crosswise, each horizontal line showing the differences to be found in the makes mentioned in that particular tire cross-section size. Thus, the D sections show the differences for $3\frac{1}{2}$-inch tires, E those for 4-inch tires, F those for $4\frac{1}{2}$- and 5-inch tires, and G those for $5\frac{1}{2}$- and 6-inch tires, rims for which are not produced by all makers.

Fig. 31. Operating Device on the Ashley-Moyer Double Q.D. Rim for Wire Wheels

Other Removable Forms. Outside of the regular range of wood wheels and the standard tires for them, any different wheel calls for a different treatment. As has already been mentioned under the subject of Wire Wheels, few of these have anything but a solid one-piece clincher rim; first, because the wheel itself is removable, thus making it as easy to change wheels as to change rims in the ordinary case; and second, to save weight and complication.

Demountable for Wire Wheels. However, demountable forms have been produced for wire wheels, one being shown in Figs. 31 and 32. This is the G-R-C double Q.D. rim as the makers prefer to call it, in action a demountable-detachable form, the clincher rim being of the straight split type, in fact, a Stanweld No. 20. This is made with a double wedging surface on the

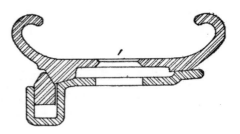

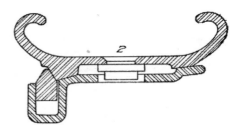

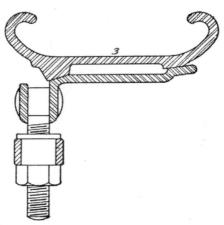

Fig. 32. Section through Rim and Band of G-R-C Rim, Showing Wedging Band and Its Operation

outside and a single one on the inside. The latter contacts with another on the false rim to which the wire spokes are attached, as does also the inner wedging surface on the outer wedge. The outer wedging surface is made so as to come just above a fairly deep slot in the false rim. In this is placed a ring with a double wedge-shaped upper edge and a square lower edge. This ring is split at one point and locked in the highest position at the point diametrically opposite.

At the split point, a pair of bent-arm levers, Fig. 31, are connected to the two ends. Attached to a middle point of each of

Fig. 33. Construction of Parker Hydraulic Steel Wheel Spokes, and Operation of Locking Device for Rims

these is one end of an inverted U-shaped member, the center and upper part of which form a bearing for a locking stud, which is attached to one end of the ring. Above this is placed a nut. As will be noted, this forms a toggle motion, the action of which is to expand the whole ring when the nut is screwed down and to contract it when the nut is screwed up.

This is the precise action used, the single ring forming the whole locking means, and being actuated by the toggle mechanism through the medium of screwing the nut up or down.

While at its best on wire wheels because of its simplicity, this rim is, of course, applicable to wood wheels. At present, its makers are specializing on the wire-wheel forms.

Parker Rim-Locking Device. Another rim-locking device which does not come under any of the standard divisions, being devised for use on the Parker hydraulic wheel is the Parker modification of the former Healy rim.

Fig. 33 shows the end of a steel spoke in section. This is made with a cup at the upper and inner end, while at the outer is a loose clip, through which passes a bolt with a head on the outside. Tightening the bolt by means of the external head draws the clip up the incline at the bottom of the cup, against the wedge on the underside of the rim, the amount of pressure exerted depending solely upon that applied to the bolt head. As the two wedge shapes oppose each other, this holds the rim as firmly as is possible. It will be noted that this construction does away altogether with the use of felloe bands or false rims used on other forms of rims or wheels, thus saving much weight. Moreover, a great part of the weight is saved at the outside, where the flywheel effect of rapid rotation is thus lessened. Moreover, the absence of additional metal here would give the tire more chance to radiate its heat, and thus would preserve it better. This construction, considering its many advantages, should have a wide use.

Similarly, with all demountable rims, the tendency is toward wider use, with which comes lower cost, as well as a better understanding of their use, abuse, attachment, and detachment. With the standardization of tires to a few standard sizes, say 9 instead of 54, it will be only a few years before all kinds of rims, including demountables, will be standardized, at which time the latter will come into universal use.

Rim Troubles. Whenever a new tire is placed on the rim, it is an excellent plan to thoroughly clean the rim and free it of dust. Sometimes the tube will come into contact with the rim, especially if the tire flap is not in position, and the sharp rusty points will chafe the tube and eventually cause a puncture. A tire can be mounted on the rim much easier if the rim is coated with a thin coating of grease and graphite. It will also keep the tire from sticking to the rim. When removing a tire from the rim of the split type, the rim should not be sprung as it will be difficult to bring the parts of the rim back into their correct position and prevent the locking device from being put into place easily.

LEFT SIDE VIEW DIANA LIGHT, STRAIGHT-EIGHT, FOUR-DOOR, STANDARD SEDAN

FORD CONSTRUCTION AND REPAIR

PART I

CONSTRUCTION OF PARTS

Introduction. In preparing this treatise on the construction and repair of the Ford automobile, the writer has constantly laid special emphasis upon the principle of operation of the various units ·and the most practical methods of repair. There is no doubt that the automobile has become one of the most widely used machines today, and necessarily it has served as an educator in mechanics. At the same time it is only wisdom—and certainly economy of both time and effort—for one to acquire his knowledge of the automobile by profiting from the experience of those who have proved their right to lay down proper practice than it is to start to do repairing with little or no knowledge of the automobile.

Familiarity with the parts and the principle of their operation makes it much easier for the mechanic or the owner to locate his troubles; if he is not acquainted with the right methods of operation or the functions of the parts, he is usually at a loss to know where he must look for the seat of the trouble.

In order to understand the operation of the different parts of the Ford car, a general description of and a detailed statement of the purpose of the various units will first be given, starting with the simple ones and progressing to the more complicated assemblies. Complete instructions for these units are given in other parts of this text where each unit is taken up in detail.

Additions to the electrical equipment on all Ford cars at the factory have opened up a hitherto unknown field to the mechanic —a field into which the Ford repair man must be equipped to enter from now on. The subject of electrical equipment has been treated very thoroughly in this volume, special emphasis being given to the methods of locating grounds, shorts, opens, etc., in the various units. Diagrams and sketches of wiring connections are also given.

Fig. 1. Plan View of the Ford Chassis

Frame. The purpose of the frame is to mount the various units in their respective order. These units are: motor, radiator, springs, steering mechanism, body, etc. The Ford frame is constructed of pressed steel, of the U section, as this type of con-

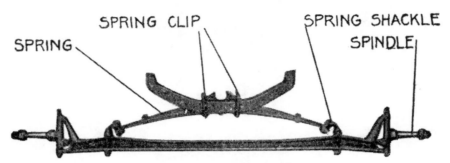

Fig. 2. Front Axle Assembly

struction gives the greatest strength for the smallest amount of steel. There are two cross members, one at each end of the frame. The purpose of these cross members is to hold the side members in place, and these members are securely fastened together with rivets. Fig. 1 is a plan view of the Ford chassis, showing the locations of the various units.

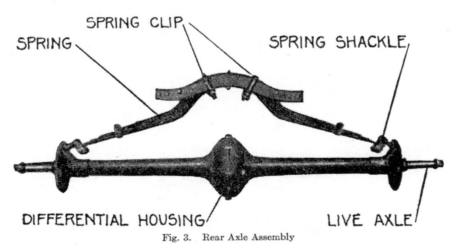

Fig. 3. Rear Axle Assembly

Front Axle. The function of the front axle, Fig. 2, is to support the weight of the front of the car through the front springs, and this axle also holds the front wheels in place by means of spindle bolts. These spindle bolts act as turning pivots for the

340

wheels when turning a corner. The front axle is drop forged in an I-beam section. Nearly all cars use a drop-forged axle as this

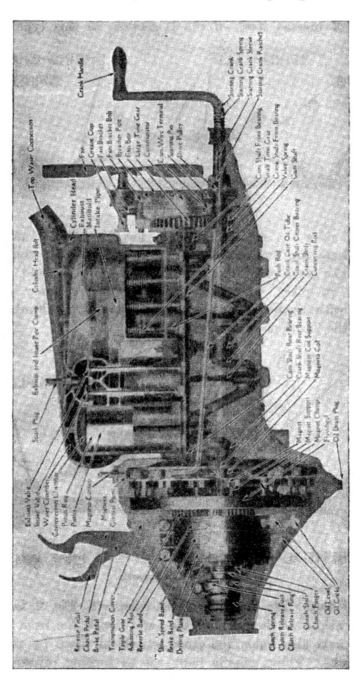

Fig. 4. Ford Unit Power Plant

construction greatly reduces the chance of flaws. It is not considered safe to use a cast-steel or a cast-iron axle.

Rear Axle. The rear-axle housing, Fig. 3, is of drawn-steel tubing. The parts are riveted together, thereby making a strong and rigid construction. These joints are also brazed to make them oil tight. The live-axle shafts and the differential gears are contained in this housing, together with suitable bearings which prevent undue wear. The driving wheels are mounted one on each rear-axle shaft. The differential equalizes the load between the two wheels.

Power Plant. The power plant, Fig. 4, is the most important unit as it transforms the gasoline into the power which drives the car. The Ford power plant is of the single-unit type as the transmission and the motor are made in one unit. A common oil case is used, although the transmission cover is a separate casting. The motor is supported on the frame at three points, this construction allowing great elasticity and preventing, to a large extent, breakage of the motor supports.

Fig. 5. Kingston Carburetor

Power-Plant Accessories. There are several necessary auxiliary systems or units that must be used with every motor. These are the carburetion, ignition, lubrication, and cooling systems.

Carburetor. The carburetor is an instrument that mixes gasoline with air in the proper proportion. The mixture is fed to the motor in variable amounts, determined by the distance the throttle is opened by the driver. Fig. 5 is a view of a Kingston Ford model carburetor.

Ignition System. The ignition system furnishes a hot electric spark in each cylinder at an exact predetermined time after the gas vapor has been compressed and is ready to be exploded. The resulting explosion creates a high pressure on the piston and drives it down, thereby turning the crankshaft. The Ford ignition system consists of a source of current—either magneto or battery—a timer, four coils, four spark plugs, and wires to make the necessary connections. A pictorial arrangement of the ignition system is shown in Fig. 6.

Lubrication System. All moving parts in any mechanism must be lubricated in order to prolong their life. In a gasoline

Fig. 6. View of the Ignition System

motor the heat of the explosion must also be taken into consideration, which makes it a particularly difficult mechanism to lubricate. The lubrication system consists of a reservoir, means of lifting the oil to a higher level than the parts to be lubricated, and a pipe to carry the oil to the gear case in the front of the motor.

Cooling System. The explosions produce a great amount of heat and cause the cylinders to become very hot. Some step must then be taken to radiate this heat; if the cylinder becomes too hot, the piston will tend to expand to a larger size than the cylinder. The piston will then stick, and there will be total failure of motor operation. The cooling system consists of a radiator and a water jacket surrounding the cylinders. The water circulates through its jacket, lowering the temperature of the cylinders to a point where proper lubrication is possible. The water then passes through the radiator, where it is cooled. The circulating system of the Ford motor operates on the thermo siphon principle. As hot water, by its physical properties, is lighter than cold water, the hot water goes to the top of the cooling system from the motor. It is cooled in the radiator and travels down by gravitation; then it enters the cylinder water jackets at the bottom. This action continues until the motor stops and the water is cold. Fig. 7 shows the cooling system.

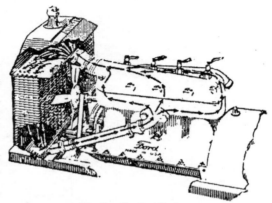

Fig. 7. Cooling System

Springs. There are two springs, of the semi-elliptic type; one mounted above the rear axle and fastened at each end to the axle housing by spring shackles and at the center to the frame by two spring clips; the other spring is mounted above the front axle and is fastened at each end to the axle by spring shackles and at the center to the frame cross member by two spring clips, as shown in Figs. 2 and 3.

Steering Mechanism. The steering mechanism is mounted on the left side of the car. (The left side of any car is always the driver's left when sitting in the driver's position.) The steering column is supported at the center by the dash and is bolted to the frame at its lower extremity. The front steering knuckles are fastened together by a connecting rod, or tie bar, and the steering column is connected to this bar by a drag link.

OPERATION OF PARTS

Motor. *Cycle.* The motor is a four-cylinder, four-cycle, L-head type. A four-cycle motor should really be spoken of as a four-stroke-cycle motor, as this type requires four complete strokes of the piston to produce an explosion; the word "stroke" is generally omitted for brevity. In other words, one power stroke is produced from each cylinder every four strokes, or two revolutions of the crankshaft. As there are four cylinders in the Ford

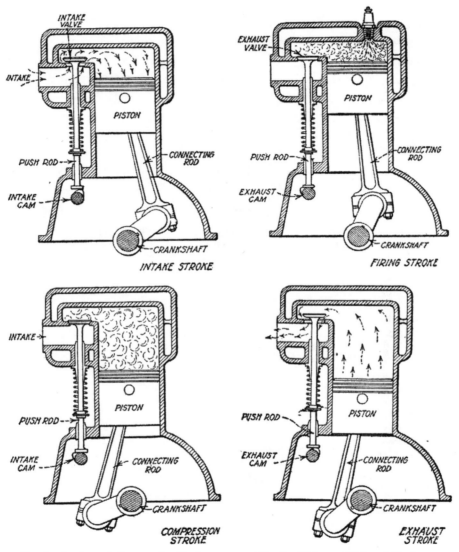

Fig. 8.   Intake and Compression Strokes   Fig. 9.   Firing and Exhaust Strokes

motor, a power stroke is produced every half-revolution of the crankshaft. The operation of the intake and the compression strokes of the cycle are shown in Fig. 8 while the firing and the exhaust strokes are shown in Fig. 9.

Intake Stroke. As the piston starts down, the inlet valve opens and a charge of gas vapor is drawn into the cylinder.

Compression Stroke. The piston then moves upward, and as both valves are closed, the charge is compressed to a pressure of about 60 pounds per square inch.

Power Stroke. An electric spark then explodes this charge, increasing the pressure about four times, or to 240 pounds per square inch. This pressure drives the piston down with great force.

Exhaust Stroke. The piston then moves upward—the exhaust valve opens before the piston reaches lower dead center—forcing the exhaust gases out of the cylinder, into the exhaust manifold, through the muffler, and into the atmosphere. This burnt gas must be expelled to make room for the fresh gas that is to be taken in at the next stroke. As both valves are on the same side of the motor, the cylinder and the combustion chamber have an inverted L shape. This is why it is called an L-head motor.

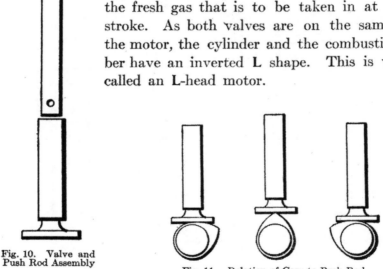

Fig. 10. Valve and Push Rod Assembly

Fig. 11. Relation of Cam to Push Rod

Valve Mechanism. The valves are made of two pieces, a cast-iron head $1\frac{1}{2}$ inches in diameter and a steel stem $\frac{5}{16}$ inch in diameter and $5\frac{1}{8}$ inches long. The stem is welded to the head with electricity. The outer edge of the head forms the valve seat and is turned and ground to an angle of 45° to fit the valve seat in the cylinder. A small hole is drilled near the end of the valve stem to allow a pin to be inserted. This pin holds the valve spring and seat in place. Operating directly under the valve stem is a push rod which is driven upward at regular intervals by a cam on the camshaft. These parts are shown in Figs. 10 and 11.

Camshaft. The camshaft is made of vanadium steel, hardened and ground on all cam and bearing surfaces to eliminate wear. The eight cams on this shaft, four intake and four exhaust, are forged integral with the shaft as this construction prevents any chance of loose cams. The camshaft and the valve assembly are shown in Fig. 12.

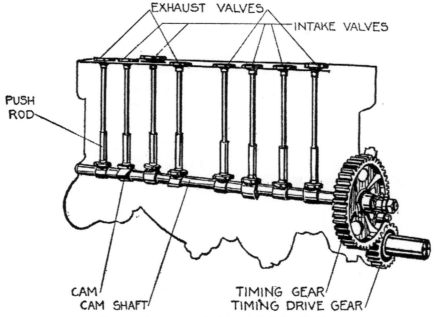

Fig. 12.   Mechanism of Valves in Relative Position

Timing Gears. There are two gears in the timing-gear case located at the front of the motor. The drive gear is mounted on the crankshaft, and has twenty-one teeth in mesh with a driven gear which is mounted on the camshaft. As the driven, or camshaft, gear has forty-two teeth, it runs one-half as fast as the motor. This speed reduction is necessary as it requires two complete revolutions to produce an explosion; therefore, the intake valve and the exhaust valve must function once in two revolutions. These parts are shown in Fig. 12.

Piston. The piston is made of cast iron and is $3\frac{3}{4}$ inches in diameter. Its purpose is twofold. The piston rings are mounted on the piston so that the compression can be securely held; it also transmits the explosive force to the crankshaft through the connecting rod. The piston must be a perfect fit in the cylinder, although a clearance must be allowed between the piston and the cylinder, this fit being secured by turning the piston and rolling the cylinder. The wristpin is mounted in the piston at each end, and the connecting rod is fastened to the center of this pin, the pin bearings being at each end of the pin. Bronze bushings are mounted in the piston and form a bearing surface for the pin. The pin is made of hardened steel, ground to the exact size, and is in the form of a tube for the purpose of reducing the reciprocating weight.

Fig. 13. Piston and Connecting Rod.

Connecting Rod. As several severe strains must be withstood by the connecting rod, therefore it is a vanadium-steel drop forging of the I section. Some manufacturers have used tubular connecting rods, but the I construction has been found to be the most satisfactory for connecting-rod use. The piston and connecting-rod assembly is shown in Fig. 13.

Crankshaft. The crankshaft has three main bearings and is of the four-throw type. The front and center main bearings are shorter than the rear main bearing, as the rear bearing has the greatest amount of strain. Nos. 1 and 4 crankpins are in the same plane, the pistons of these cylinders traveling together; 2 and 3 are likewise in the same plane. The crankshaft is also a vanadium-steel drop forging, all bearings and crankpins being accurately

Fig. 14. Ford Crankshaft

turned and ground. The crankshaft is shown in Fig. 14; the end thrust of the crankshaft is taken up by the rear main bearing.

Flywheel. The flywheel is bolted to a flange at the rear of the crankshaft. A flywheel must be used to store up energy between the high and the low peaks of the power strokes. This energy is given off when there is but little energy being produced, this time being between the opening of the exhaust valve of one cylinder and the starting of the power stroke in the next cylinder to fire. It would be impossible to run the motor very slowly unless the flywheel was used.

Magnets. There are sixteen magnets mounted on the flywheel. These magnets are used in conjunction with sixteen coils to generate an electric current for ignition. In the early Ford models the lights were also supplied with current from this magneto. A complete description of this instrument will be found in the section on the "Electrical System," Part II.

Transmission. It is a generally known fact that transmissions are not used on steam cars or electric vehicles. The question then arises, why is a transmission—or change-speed gears—necessary in the gasoline automobile? To make this clear, it will be necessary to call to mind a few principles of the internalcombustion motor.

The Why of Transmission. The steam engine is an external-combustion motor; in other words, the fuel is burned in the firebox of the boiler, outside the cylinder. The steam pressure is generated in the boiler, and the steam is conveyed to the cylinder through a pipe. When the throttle is opened, the steam pressure forces the piston down. If there is a heavy load on the engine, preventing it from turning, the pressure will still be maintained until the boiler pressure is exhausted, the hot steam from the boiler supplying the heat that is absorbed by the cylinder. When the piston moves down a certain distance, the steam supply is shut off, this distance depending on the point of the valve cut-off. The steam contained in the cylinder then expands and forces the piston out the remainder of the stroke.

The action in the gasoline engine is very different from that in the steam engine as the gasoline engine is an internal-combustion motor; that is, the fuel, in the form of a gas, is taken into the cylinder as the piston travels down on its intake stroke. The inlet valve then closes and the gas is compressed as the piston travels toward the combustion chamber. An electric spark then ignites this gas and raises the temperature to about four times that of the compression temperature. As the pressure is in direct proportion to the temperature, it is also increased about four times. Now, suppose that the motor is connected to a heavy load, as in the steam engine. The explosion is not strong enough to move the piston and the heat is quickly absorbed by the cylinder. This loss of heat lowers the pressure in the same pro-

Fig. 15. Section of the Ford Transmission

portion that it was originally raised, and since no outside heat can be added as in the steam engine, the pressure continues to decrease until it reaches zero.

All this heat which has been absorbed by the cylinders is wasted heat, or energy—all combustion motors are heat engines— and additional heat must be supplied in order to furnish constant pressure on the piston and produce a constant turning force, or turning torque, on the crankshaft. As this heat cannot be continuously supplied, it is necessary to use transmission reducing gears. This allows the motor to turn faster than the drive shaft, thus increasing the number of explosions per revolution of the drive shaft. This gives a greater turning torque to the motor and enables it to start the car under a heavy load and to climb steep hills when loaded. The transmission thus transforms the internal-combustion motor into a power plant which has an elasticity nearly approaching that of the steam engine.

Ford Transmission. The Ford has a planetary-type transmission which is entirely different from the transmission used on other pleasure cars. Fig. 15 shows the assembly of the Ford

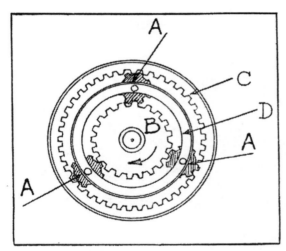

Fig. 16. Principle of the Ford Transmission

transmission, while Fig. 16 is a simple diagrammatic sketch showing the principle of its operation. The gear *C*, however, is eliminated in the Ford car and a small external gear is used instead. This does not alter the principle of operation. Let us suppose that the gear *B*, which is continually in mesh with the gears *A*, is mounted on the crankshaft of the motor. This gear will revolve in a clockwise direction as indicated by the arrow. The three gears *A* are mounted on the common spider *D*, and they are continually in mesh with the external gear *B* and the internal gear *C*. In order to obtain low speed, the gear *C* must be locked. The drive shaft is connected to the spider *D*, and this spider will revolve in a clockwise direction. As *B* is turning *A* in a clockwise direction, and as *C* is locked, the gears *A* will travel within *C* in a clockwise direction, or in other words, the spider *D* will revolve clockwise. When it becomes necessary to reverse the direction of the car, the gear *C* is released and connected to the drive shaft and *D* is locked. *B* will then cause the gears *A* to revolve on their spindles in an anti-clockwise direction, and as *D* is locked, the internal gear *C* will revolve in an anti-clockwise direction. When high speed is used, a separate disc clutch is operated which connects the motor crankshaft to the drive shaft. The entire arrangement then revolves as a unit, which has the effect of a flywheel.

Clutch. It is necessary to have some means of connecting and disconnecting the power of the motor from the rear wheels in order that the motor may be started and that the car may be put in motion without stalling the motor or causing the car to jerk.

Fig. 17.   Construction of the Rear Axle

Fig. 35 shows the clutch parts. The clutch performing this function is of the disc type.

The clutch consists of a number of steel discs, half of them being fastened to the motor and the other half fastened to the drive shaft. The motor discs are set alternately between the drive-shaft discs. These discs are normally held together by a coil spring, thus causing the drive shaft and the motor to turn as a single unit. When the driver wishes to disconnect the power of the motor from the drive shaft, it is only necessary to release the tension of this spring to allow the discs to slip between each other. This is accomplished by the driver's pressing on the clutch pedal with his left foot part way or by drawing up the control lever.

Rear=Axle Assembly. The drive shaft connects the clutch to the rear axle. As the shafts are mounted at right angles, it is necessary to transmit the power through a set of bevel gears. The standard ratio between these units is $3\frac{7}{11}$ to 1. There are two axle shafts connected at the center by the differential frame; the differential-gear arrangement, Fig. 17, shows the construction of the rear axle.

Differential. One of the most important mechanisms of any automobile is the differential, but as this unit causes very little trouble, few laymen are aware of its presence or realize its impor-

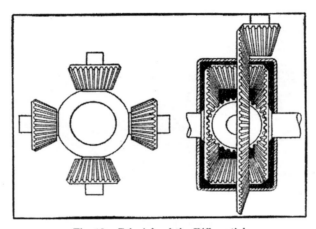

Fig. 18. Principle of the Differential

tance. It would be difficult to control the car without a differential when driving around corners since, when turning a corner, the outer wheel must turn faster than the inner one as the outer wheel is describing a larger circle.

The differential gear is used so that either wheel will be allowed to turn faster than the other and at the same time equalize the power transmitted to each wheel when driving straightaway. Fig. 18 shows the principle of the differential. Here it will be seen that the axle is divided into two distinct axle shafts. A bevel gear is carried at the inner end of each shaft, and these gears are keyed to the axles so that they revolve with them. Intermediate bevel gears are mounted on a spider and these gears mesh continually with the axle gears. The ring gear is bolted to the housing, this housing turning as a unit with the intermediate gear spindles. The intermediate gears are free to turn on their spindles and do so when the resistance of one rear wheel is greater than that of the other. This resistance tends to hold the inner gear and prevent it from turning as rapidly as the other axle gear. The intermediate, or differential, gears then turn on their spindles, thus equalizing the power on the rear wheels. Now let us assume that all gears are in mesh, that power is being applied to the ring gear, and that the resistance is the same at both rear wheels. Then the entire assembly—comprising ring gear, differential, intermediate, or pinion, gears, and axle-shaft gears—revolves. If both wheels are turning forward at the same speed,

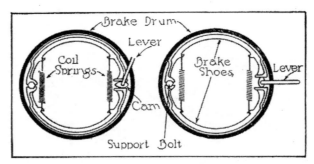

Fig. 19.   Emergency Brake

the differential pinions remain stationary and merely act as a lock, forming a driving connection between the driving gears. This will cause both wheels to turn in the same direction as long as the load is uniformly distributed.

Brakes. The braking system forms a very important unit in the control of the car. If the motive power is disconnected by the clutch, the car will continue to move on account of its momentum, and it is therefore imperative that some means of stopping the car be provided.

There are two braking systems on the Ford car, one operated by a foot pedal, the other by the hand lever. The foot, or service, brake consists of a brake band operated on a brake drum in the transmission unit. When the foot pedal is depressed, this band is tightened, causing the speed of the transmission brake drum to be checked. As this brake drum revolves at the same speed as the drive shaft, the speed of the car is materially lessened. This brake is the one to use when driving the car and is operated by the right-hand foot pedal.

The emergency brake is of the type shown in Fig. 19, where there is a brake for each wheel operating on the inside of the brake drum. The brakes used on the Ford models of a few years ago were of cast iron made in the form of two semicircular shoes. The brakes now used are made of cast iron in the form of a

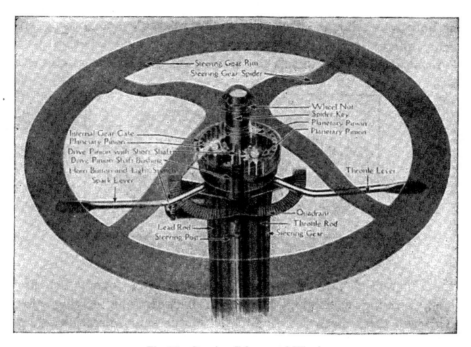

Fig. 20.   Steering Column and Wheel

circular shoe. But the center portion of the shoe at the back is made very thin, so that there is considerable give without breaking. The shoe is split at the front to allow a flat cam to be placed between the ends of the brake shoe. Two coil springs are fastened between the brake shoes to hold them in the released position when the cam is flat. If, however, the cam is rocked, so that instead of lying flat it is moved to such an angle as to cause the brake shoe to spread, it will grip the internal surface of the brake drum and retard the movement of the wheels or entirely stop the movement of the car. The braking effect depends on the distance the brake arm is moved by the driver. This brake is intended to be used only when the car is standing or it is necessary to lock the rear wheels when on a hill.

Steering Gear. Every car must have some suitable means of steering. The steering mechanism, however, must be so constructed as to give unfailing service without undue strain on the driver.

The steering, or wheel, post is a metal rod carried inside the steering column which is capable of being turned a certain number of degrees so that the steering arm connected to the drag link may be moved. This drag link is connected to the connecting

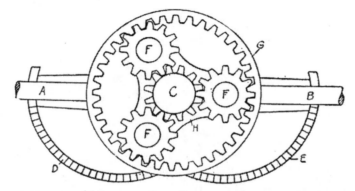

Gears in Steering Column: A, Spark Lever; B, Throttle Lever; C, Pinion on End of Steering Wheel Shaft; D, E, Quadrants; F, Spider Pinion Gears; G, Internal Gear.

Fig. 21.   Steering Reduction Gears

rod which joins the steering knuckles. The turning of the steering wheel is thus transmitted to the front wheels in such a manner that the car may be satisfactorily steered. The limit of movement is determined by the distance the front wheels can be moved. Fig. 20 shows the construction of the steering column. The steering post is housed in a metal tube of sufficient size to carry the spark and the throttle control rods. These rods are worked by levers placed below the steering wheel and convenient to the driver's reach. The steering column is set at such an angle that the steering wheel is in a convenient position to the driver. The steering wheel consists of a metal spider having four arms terminating at the oval wooden rim: the intersection of these arms at the center forms a boss. A hole is machined to allow the end of the steering post to enter and form a tight fit. A key prevents the wheel from turning on the steering post, and a nut screwed on the end of the steering post holds the wheel in place.

Ford Steering Gear. The Ford steering gear differs radically from the form used on the conventional car. The spider pinion gears, which permit a greater movement of the hand than of the steering arm, are located at the top of the steering post instead of at the bottom as in other cars. Instead of using the worm gear—the form most used in the average car—a planetary-gear arrangement is employed. These gears are in a compartment below the steering wheel and are packed with a lubricant to ensure perfect operation. The construction of this mechanism is shown in Fig. 21. This case contains four spur gears, one in the center, with three surrounding it. These three gears F are each mounted on individual studs, or spindles, the spindles being attached to a common triangular plate (or spider) H connected to the top of the steering post.

The casing G is provided with teeth on its inner periphery, and the three spur gears are in continual mesh with this internal gear. When the steering wheel is turned, the center spur gear C, connected to the steering wheel, turns the three gears, forcing them to travel around the inner periphery of the housing. This turns the triangular plate in the same direction but much slower. The driver is then enabled to handle the car very readily, even on unfavorable roads.

OVERHAULING THE CAR

Cleaning Car. One of the first steps in the actual work of overhauling the power plant is to clean the car very thoroughly. Any dirt or grit allowed to get into the bearings is likely to cause trouble by cutting the bearings when the car is again assembled. The gaskets must also be kept clean, although this is very hard to do if the adjacent surfaces are covered with dirt and grit. If particles of dirt get on any of the gaskets, it will be difficult to make a compression-tight joint. It is generally advisable to use new ones.

Identification of Parts. While the skilled mechanic is supposed to know where each and every part of the Ford car belongs, still the individual car owner or the apprentice mechanic is not usually so proficient, and a great deal of trouble will be avoided by marking the parts as they are removed from the car.

One method of marking is by small tags or by pieces of paper through which the bolts are forced. Particular care should be taken in marking those parts the use of which is not obvious. Another good method of identification is to number the parts with the same numbers that they carry in the Ford parts list. Such parts as the eight valves can be marked with a center punch, beginning at the front of the motor with one dot and proceeding toward the rear of the motor; the last valve will be marked with eight dots.

In using a center punch for marking the valves, the valves should be marked before removing to prevent bending the valve stem. Such parts as the gaskets for the radiator and the two screws holding the cylinder-outlet water-hose connection to the cylinder head can be tied on the radiator with a string. This method of tying the gaskets and other small parts to the larger parts to which they belong is a great time saver when the car is assembled. The use of cigar boxes, or boxes of miscellaneous sizes, in which to keep the parts of the different elements of the car is also helpful. For instance, all motor parts should be placed in one box, while the transmission cover bolts and other small parts belonging to the transmission should be placed in another box.

Removing Radiator. The radiator should be drained preparatory to removing it from the car. Removing the radiator makes it easier to get at the engine, commutator, and other parts. In order to drain the radiator, it is usually necessary to clean out the radiator drain cock after the cock has been opened, as mud and sediment usually accumulate in the bottom of the radiator. Fig. 7 shows the radiator and the hose connections.

The radiator is supported by a bracket at each side and by the water inlet manifold, with a truss rod between the dash and the radiator. This bracket is bolted to the side members of the chassis frame. The inlet manifold of the radiator is located in the center of the base of the radiator header tank, and the outlet manifold is offset at an angle from the left side of the bottom tank, or base of the radiator. The inlet manifold is connected by a rubber hose and a flanged fitting bolted to the outlet of the water jacket at the forward end of the cylinder head. The outlet manifold is bolted to the intake of the water jacket at the left side of the cylinder block near the base of the water jacket. The radiator outlet manifold consists of a metal tube and a hose at each end; there is also a flanged fitting fastened to the cylinder block and connecting with this manifold. Two cap screws secure the flanges of the fitting at the outlet and at the inlet on the cylinder head.

OIL DRAIN PLUG

OIL PAN REMOVED

Fig. 22. Bottom of the Crankcase Showing Drain Plug

While the radiator is removed from the car, the radiator hose connections should be taken off and inspected. No doubt it will be advisable to put on three new hose connections at this time since this will tend to eliminate water leaks or clogged water hose at some future time. Sometimes new hose clamps are also needed. It is also a good plan to adjust the fan belt at this time, and if it is badly worn, it should be replaced with a new one. For regular use, either the leather or the fabric belt gives good results, and a belt with a coupling can also be carried in the car for emergencies.

Draining Oil. We are now ready to drain the old oil from the crankcase. After some use, the oil becomes black and dirty and is filled with metallic particles that have been worn from the bearings and other parts. Also, after the oil has been used for some time it is much thinned out, since the kerosene condensed in the combustion chamber works down past the piston rings and destroys the lubricating qualities of the oil. Much wear and tear will be saved if the oil is changed about every thousand miles.

After removing the crankcase drain plug, located below the flywheel, and draining out the old oil, about 2 quarts of kerosene should be poured into the crankcase at the filler spout. The position of the drain plug is shown in Fig. 22. Then the motor should be turned by hand for several revolutions, splashing the kerosene around in the crankcase to clean out the old oil. If the front end of the car is jacked up about 6 inches, the kerosene will run back into the oil reservoir and drain out more completely. The drain plug should then be replaced. The old oil drained from the motor should not be used again in the crankcase; it is, however, satisfactory for spring lubrication and other minor places on the chassis.

OVERHAULING MOTOR

Preliminary Operations. The first part to be removed is the cylinder head; but before doing this, it is advisable to remove the four spark plugs to prevent them from being broken. Fig. 23 is a view of the motor showing the head detached.

The fifteen cylinder-head bolts should be removed with the socket end of the spark-plug wrench or a socket speed wrench, which can be purchased from any supply house. Less time is required when using the speed wrench.

Before removing the rear cylinder bolts, it is necessary to take off the small metal plate on the dash under the coil box.

If the valves are to be ground, it is advisable to remove the exhaust manifold, and the carburetor and intake manifold in one unit; this will permit easy access to the valves.

Before removing the carburetor, it is first necessary to disconnect the gasoline pipe. The gasoline supply should be shut off at the sediment bulb under the gasoline tank. It will also be necessary to lift the carburetor adjusting rod located on the dash.

In removing the exhaust manifold, it is easier to take out the exhaust pipe and manifold in a single unit by a straight pull forward. When removing the manifolds, be careful not to injure the gaskets, as these gaskets ensure a gas-tight joint between the manifolds and the cylinder block.

Removing Valves. The first step when removing the valves is to remove the valve-chamber cover plates, after which the valve springs may be compressed by means of a suitable valve-lifting tool and the valve pins pulled out. These pins should be placed where they will not be lost. Two of the valves are always in the raised position, and it will be necessary to turn the crank until these valves go down, when the springs can be compressed and the valve pins removed.

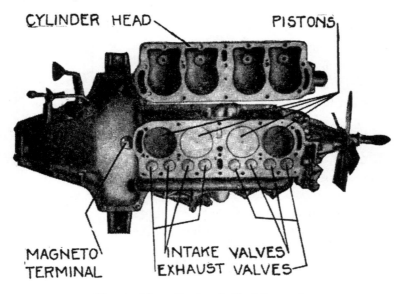

CYLINDER HEAD PISTONS

MAGNETO TERMINAL INTAKE VALVES EXHAUST VALVES

Fig. 23. Motor Showing the Head Removed

Crankcase Repairs. We are now ready to remove the lower cover door of the crankcase. To do this, it is necessary to get under the car and remove the fourteen $\frac{5}{16}$-inch cap screws. A special speed wrench is made for spinning out these cap screws. This wrench, Fig. 24, is very short so that it can be used in the limited space. In removing the lower crankcase door, the gasket generally sticks and should be renewed. There is always some oil left in the connecting-rod dip pans on this cover, consequently one should not be directly under the pan when it is removed.

Adjusting Connecting Rods. We are now ready to tighten the connecting-rod bearings, which is best done one at a time.

Begin at the front connecting-rod bearing and examine the bearing cap to be sure that it is marked. There is a file mark on this cap on the side toward the camshaft, and if this bearing cap is not so marked, this should be done before the cap is removed. After removing the connecting-rod cap, file off a small amount of metal or remove one or more shims, if there are any. Then replace the cap and tighten the bolts, but do not replace the cotter pins at this time.

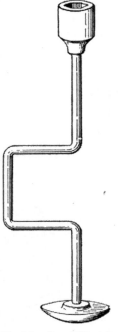

Now try the tightness of the front connecting-rod cap by turning the starting crank. It should be possible to turn the crank, since the bearing should be a snug fit only.

After tightening the front connecting-rod cap, loosen the nuts on the bolts a couple of turns, then proceed in the same manner with the second connecting rod. If the cap is too tight when the bolts are securely tightened, place one or more shims between the connecting rod and the cap so that the engine will not be too hard to crank when all four bearings have been tightened.

Fig. 24.   Speed Wrench for Inspection Plate

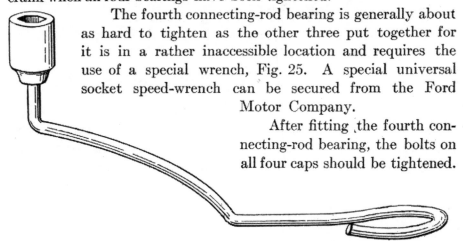

The fourth connecting-rod bearing is generally about as hard to tighten as the other three put together for it is in a rather inaccessible location and requires the use of a special wrench, Fig. 25. A special universal socket speed-wrench can be secured from the Ford Motor Company.

After fitting the fourth connecting-rod bearing, the bolts on all four caps should be tightened.

Fig. 25.   Fourth Connecting-rod Wrench

Then the cotter pins should be put in and the ends of the cotter pins spread to keep them from dropping out.

Caution. Care should be taken not to get any broken ends of the cotter pins in the crankcase. Such bits of metal might be carried by the oil onto the magneto coils and cause a short circuit which might result in total failure of motor operation. A rag should be placed between the coil support and the crankcase.

Piston Slap. If there has been piston slap in the motor, new and oversize pistons should be installed at this time. It is sometimes advisable to rebore the cylinders when they are badly worn. This can be done in almost any good machine shop at small expense, or the small garage can do the job with a reboring tool made especially for that purpose. There are several of these

CARBON DEPOSIT.

Fig. 26.   Removing the Carbon

reboring tools on the market, and they may be purchased at a supply house. If the original pistons which came with the car are still in the engine, then the new pistons should be .0025 inch oversize. However, if the cylinder block has been rebored and is fitted with 0.03125-inch oversize pistons, then pistons .033-inch oversize should be installed.

Adjusting Main Bearings. Where there have been main-bearing knocks—they usually cause a deep heavy thud when the throttle is open and the motor is pulling hard—it will probably be necessary to remove the engine from the car and tighten the main bearings. This is best done when the motor is removed from the frame, thus making all of the bearings accessible.

While it is not difficult to tighten the middle main bearing through the crankcase-cover lower door without removing the motor from the frame, still the middle main bearing does not give trouble very often. It is the rear main bearing which is the most frequent offender.

The rear main bearing carries the load of the flywheel and magnets and the fore part of the transmission, in addition to the load due to the force from the connecting rods. So, in spite of the fact that this main bearing is made longer than the others, it has so much additional work to perform that it wears more rapidly and is generally the first of the three main bearings to give trouble. The front main bearing can be tightened without taking the motor out of the car if there happen to be some shims between the bearing cap and the cylinder block. It is only necessary to loosen the bolts, holding the front bearing cap in place, and pull out one or more shims and then tighten the bolts.

Fig. 27. Compressing the Valve Spring

Grinding Valves. In overhauling the motor, grinding the valves and removing the carbon are generally two of the most important details that must be undertaken. The carbon is easily scraped off the piston tops and the cylinder block by a putty knife or other flexible flat-bladed tool, Fig. 26. A steel scratch brush, such as is used to scrape sand from steel castings, is also useful for removing the carbon from the Ford cylinder head.

Before grinding the valves, it will be necessary to remove all the valve springs. This is best done by the use of some good valve lifter, several of which are on the market. The lifter holds the spring compressed, Fig. 27, while the pin is removed from the end of the valve stem.

The work of grinding the valves will be made much easier if the valves are refaced with a valve-refacing tool; if the valve seat is reamed with a valve tool, then a still better job can be had in much less time. These operations are shown in Figs. 28 and 29. After the carbon has been removed from the valves, a small amount of grinding compound should be placed on the edge of the valve where the seat is formed. A light push spring about 2 inches long and $\frac{3}{8}$ inch in diameter should then be placed on the valve stem, so that, when grinding, the valve will be lifted from its seat by this spring when the pressure is removed. This action is necessary to change the position of the cutting compound

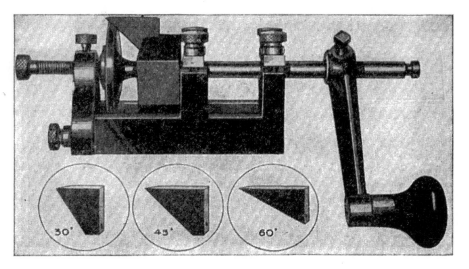

Fig. 28. Refacing the Valves

and to prevent the valve seat from being grooved. The valve should be turned with a forked tool, Fig. 30; do not turn the valve in one direction only as this will groove the seat. The spring should then be allowed to lift the valve, and when it is lifted, it should be turned so that the position of the grinding compound will be changed. This compound is sold in a box with two compartments, one containing the coarse and one the fine compound. The fine compound is generally satisfactory for the intake valves.

After grinding the valves, great care should be taken to clean out all the grinding compound. Do not allow the compound to get on the pistons or the cylinder walls as it will cause a great deal of wear if left on these parts.

Inlet Valves. In grinding the valves, it will be noticed that the inlet valves are not usually pitted and scored as much as the exhaust valves. The reason is that the exhaust valves are subjected to the hot flame of the exhaust gases, while the inlet valves are cooled by the fresh incoming gases from the carburetor. For this reason the exhaust valves become much hotter

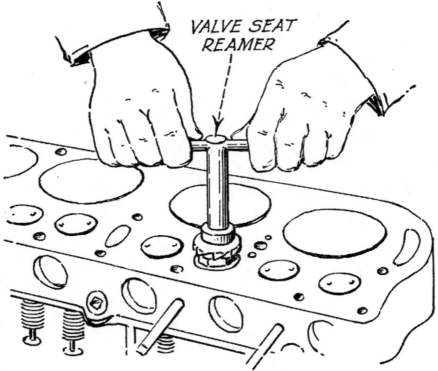

Fig. 29.   Refacing the Valve Seats

than the intake valves. This heat usually burns the carbon from the exhaust valves, while the tops of the intake valves are covered with carbon. As the inlet valves are generally in fair condition, it is not necessary to reface them to the same extent as the exhaust valves.

If the motor has been in use for several years, it sometimes happens that there is sufficient wear around the stems of the inlet valves to cause air leaks at these points, Fig. 31. This air

leak will sometimes allow enough air to pass into the intake manifold to cause the motor to miss at low throttle openings; it will
also make starting difficult. Replacing the valves will reduce this
leakage to a certain extent, but sometimes the valve guides are
worn and replacing the valves is not sufficient. Then it will be
necessary to install valves having $\frac{1}{64}$-inch oversize stems. When
valves having oversize stems are used, ream out the valve-stem
guides with a $\frac{1}{64}$-inch oversize valve-guide reamer.

Adjusting Valves. After replacing the valves, it is necessary
to adjust the valve-tappet clearance, which should be between $\frac{1}{64}$

Fig. 30. Grinding the Valves

and $\frac{1}{32}$ inch. For passenger-car use, where one desires to obtain
a quiet-running motor, less clearance than $\frac{1}{64}$ inch can sometimes
be given. This tends to eliminate valve-tappet noises and clicks,
but it will also cut down the power of the motor to some extent.
Not less than 0.008 inch or 0.010 inch clearance for the inlet
valves and not less than 0.015 inch clearance for the exhaust
valves should be allowed. No adjustment has been provided by
the Ford Motor Company, so if it is desired to make this adjustment a set of valve-adjusting discs should be purchased. If the
stems are too long, they may be shortened by filing.

It sometimes happens, after the reground valves have been run in the engine for a short time, that the valves seat themselves more deeply into the cylinder block, thus reducing the clearance between the ends of the valve stems and the tappets. For this reason, it is better not to replace the covers of the valve chambers

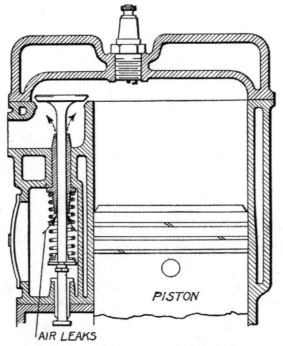

Fig. 31. Air Leaks around the Valve Stems

but to measure the valve-tappet clearance and adjust the valves again after the engine has been running for fifteen or twenty minutes.

The clearance should be checked with a thickness, or feeler, gage, consisting of a number of thin strips of steel. If this tool is not at hand, an approximate valve-tappet clearance adjustment can be secured by using a postal card as a thickness gage. Such a card is about 0.010 inch in thickness.

The valves are $1\frac{1}{2}$ inches in diameter at the head, having a seat at an angle of 45 degrees to the stem. The valve stem is $\frac{5}{16}$ inch in diameter and $5\frac{1}{8}$ inches long. There is a pinhole $\frac{5}{8}$ inch from the end of the valve stem, this hole being $\frac{3}{32}$ inch in diameter. The exhaust and the inlet valves are of the same size and they are interchangeable.

Valve Timing. The valves should be timed by the position of the piston. The measurements for the different models are as follows:

Models previous to 1913

    Exhaust opens $\frac{3}{8}$ inch before lower dead center
    Exhaust closes $\frac{1}{64}$ inch past upper dead center
    Intake opens $\frac{7}{64}$ inch past upper dead center
    Intake closes $\frac{3}{8}$ inch past lower dead center

Models later than 1913

    Exhaust opens $\frac{5}{16}$ inch before lower dead center
    Exhaust closes upper dead center
    Intake opens $\frac{1}{16}$ inch past upper dead center
    Intake closes $\frac{9}{16}$ inch past lower dead center

Assembling Motor. After cleaning off the carbon and grinding the valves, the cylinder head should be replaced. But before doing this, the cylinder-head gasket should be carefully cleaned. It is not necessary to buy a new gasket if the old one has been re moved with reasonable care and is not torn or broken.

When replacing the cylinder head, turn the starting crank so that the first and fourth pistons are in the extreme raised position —on upper dead center—and are projecting above the cylinder block. The pistons will hold the gasket and keep it from slipping when in this position. Smearing both sides of the cylinder head with heavy grease not only helps to keep the gasket in position, but it will also assist in making a compression-tight joint, as the grease allows the gasket to work to the correct position and fit smoothly between the cylinder head and the cylinder block.

It is also a good plan before replacing the cylinder head to clean the carbon and dirt out of the cylinder bolt holes, using a twist drill about $\frac{3}{8}$ inch in diameter. If this dirt is not cleaned out of the bottom of these holes, it will be impossible to tighten the bolts enough to make a compression-tight joint between the cylinder head and the motor block. The bolts are also likely to be twisted off when they are being tightened. After replacing the cylinder-head bolts, spin them down with a speed wrench and then tighten them with the cylinder-head and spark-plug wrench. It is not necessary to tighten these bolts with excessive force, as they may be broken if this is done. There is a certain knack in tightening these bolts. The center bolts should first be tightened and then the bolts on each side, working toward the ends of the cylinder heads, until all the bolts have been tightened. They should then be gone over for a final tightening. The bolts should be tightened in the order shown in Fig. 32; if they are first tightened

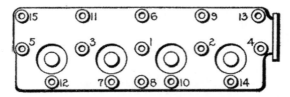

Fig. 32. Tightening Cylinder Head Bolts

at one side or at one end of the cylinder head, a compression-tight joint is hard to make and leaks may occur.

Inspecting Spark Plugs. We are now ready to inspect the spark plugs, which should be taken apart and cleaned before being replaced in the cylinder head. Emery cloth should not be used to clean the porcelains as it will remove the glaze and allow oil to soak into the porcelain; then the spark plug will short-circuit very easily. After assembling the porcelain and the body of the plug, care should be taken not to make the nut too tight, as the heat of the engine may crack the porcelain. Adjust the gap between the spark-plug points to $\frac{1}{32}$ inch and bend the grounded point of the electrode upward; then, if oil collects on the points, it will not bridge the gap but will run off to the side where it can do no harm. This adjustment is shown in Fig. 33.

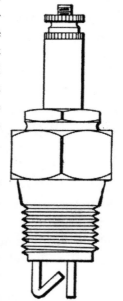

Fig. 33. Spark Plug Adjustment

Wiring. The timer wires should be replaced if they are oil soaked and badly worn or if the insulation is broken near the commutator. When examining the timer wires, stray ends should be carefully looked for, as these fine ends may touch the motor when the spark lever is moved and cause the motor to miss. The commutator wires require much more attention than the high tension cables that lead to the spark plugs. The color of the wires indicates the terminal to which they should be attached.

Timer. The timer should be taken apart and well cleaned, and if the raceway on the inside of the timer shell is worn or rough, this shell should be replaced. A worn timer shell will cause the roller to bounce and this, in turn, may cause misfiring of the engine at speeds of 25 m.p.h. or over. The timer roller assembly is another small part which sometimes causes much trouble. If the timer roller is worn, or the timer spring weak or broken, it is advisable to replace the entire roller-brush assembly.

Coil Adjustments. Another part of the ignition system upon which much of the smooth running of the motor depends is the

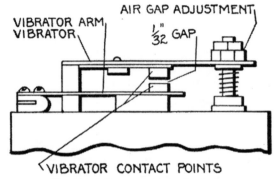

Fig. 34. Adjustment of the Spark Coils

adjustment of the vibrator points of the spark coils. These coil points should be ground smooth and true, so that they make good contact. After the coil points have been worn down about halfway, it is usually necessary to replace them with new ones in order to get an effective coil-point adjustment.

If possible, the coils should be taken to the nearest Ford agency where there is a coil-testing machine and adjusted until each unit consumes from 1.2 to 1.4 amperes as indicated by the ammeter of the coil-testing machine. The coil points should separate about $\frac{1}{32}$ inch when the vibrator is pressed down against the core of the coil unit. Fig. 34 shows the adjustment of this unit.

Preparing Motor to Run. After having made sure that the crankcase drain plug has been tightened and the crankcase lower door replaced, a gallon of clean oil should be poured into the crankcase.

We are now ready to replace the radiator and to fill it with water. After filling the radiator, the joints around the radiator hose connections and between the cylinder head and the cylinder block should be examined to see whether there are any leaks.

The engine can now be run as slowly as possible for five or ten minutes until the oil is worked into all the moving parts and the parts which have been replaced have had a chance to adjust themselves to each other. If new piston rings have been installed or if the connecting-rod bearings have been tightened, the engine should be run for an hour or so at a slow speed but with plenty of oil to give these parts a chance to work into good running condition.

OVERHAULING TRANSMISSION

Noisy Transmission. After a car has been run for several thousand miles, especially in hilly country, the transmission becomes very noisy and will grind when either the low-speed or the reverse-speed brake bands are operated. As there is no power transmitted through the gears when the car is running in high speed, little trouble will be had with the gears when the car is being driven in high. The discs in the clutch assembly may be roughened or worn enough so that they cannot be adjusted any further by means of the clutch-adjusting screws. When the power plant is out of the chassis, it is advisable to examine the transmission gearing if unusual noises have been present. The gears should not wear very much, although the bushings in these gears are subjected to considerable wear.

Tearing Down Transmission. *Clutch-Disc Assembly.* In taking the transmission apart, it is first necessary to drive out the clutch spring and the thrust ring support pin, which makes it possible to remove the clutch shift collar, *Group 5*, Fig. 35. The driving plate can be removed after the screws bolted to the brake drum are taken out. The clutch-disc assembly is then exposed as shown in *Group 4*. The clutch discs are carried by the disc drum as shown in *Group 1*. A set screw holds this member securely to the rear end of the crankshaft. After this set screw is loosened, the disc drum may be removed and the assembly will then appear as in *Group 3*.

Drum Assembly. The remaining part of the assembly, consisting of reverse drum, low-speed drum, brake drum, and triple gears, may be easily withdrawn from the flywheel and the crank-

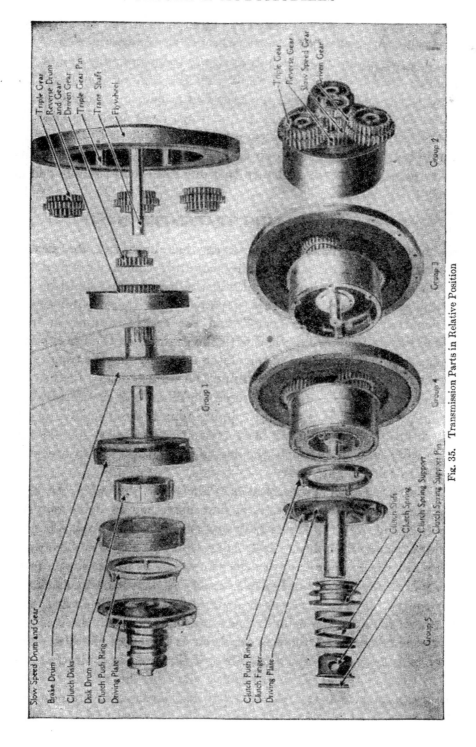

Fig. 35. Transmission Parts in Relative Position

shaft extension known as the transmission shaft. The assembly is then as shown in *Group 2*. In order to take down the remainder of the assembly, the driven gear must be removed from the brake-shaft extension, allowing the three drums to be pulled apart. The bushings in the triple-gear assembly must be examined after the transmission has been taken apart. It is also necessary to examine the pins attached to the flywheel that support the triple gears. If either the bushings or the pins are worn, there will be considerable play and the transmission will be very noisy when operating in low or reverse speed. If the bushings are worn, they should be removed and new ones installed, special care being taken to see that the new bushings are reamed concentric so that the gears will revolve true. The pins mounted on the flywheel should also be replaced if worn. The bushings in the reverse drum and gear and in the interior of the low-speed drum and gear should be carefully examined to make sure that the low-speed drum is a good fit on the sleeve of the brake drum; the reverse drum should also fit properly on the extension of the low-speed drum. If these bushings are worn so that it is considered advisable to install new ones, they should be removed and new ones forced in place by means of an arbor press or a vise.

Before reassembling the brake-drum unit, the bushings should be fitted to turn freely on the members by which they are supported. The surfaces of the brake, the low-speed, and the reverse drums should not be cut or scored. This scoring often happens when transmission bands are riveted in place with iron or steel rivets. Soft copper or brass rivets should always be used for this purpose, and the rivets should be properly countersunk.

Clutch Discs. The clutch discs should be removed, thoroughly cleaned, and inspected to see if there are any rough surfaces. If there are ridges on the plates which come together—a result of the operator continually slipping the clutch—the ridges should be removed with a file; if this thins down the plates too much, new plates should be used. It is advisable to install new plates if they are rough, as this also indicates that the plates are soft.

Transmission Band Linings. The transmission and brake lining is $\frac{3}{16}$ inch thick, $1\frac{1}{8}$ inches wide and 23 inches long. Three of these strips are required, making a total of 69 inches.

Transmission Repairs. Dismantle and clean all parts. See that all magnet clamps are tight and that magnets are parallel. Try the triple gear shafts for looseness in the flywheel; if loose, replace them with oversized shafts. Examine the triple gears for worn or loose rivets; if the rivets are not tight, peen them; if very loose, they should be replaced.

Rebushing. Try the triple gears on the shafts; if there is over .005 inch play in the bushings, they should be rebushed. When rebushed, the flange face of the new bushings should not project over .005 inch to .007 inch from the side of the triple gears. Examine the lugs on the inside of the brake drum; if they are worn or cut over $\frac{1}{32}$ inch deep on both contact sides, the drum should be scraped. If the driven-gear sleeve flange-bushing face is badly worn or too thin, it should be replaced.

Bushing Clearance. The gear shaft should be fitted to the driven-sleeve bushings to a clearance of .003 inch on a new job and on a repair job, a clearance of .005 inch. Examine the rivets on the slow-speed drum, making sure that they are tight. Also inspect the gears for worn or chipped teeth and test the clearance of the slow-speed drum on the driven-gear sleeve. There should be a clearance of .003 inch on a new job and .005 on a repair job.

Reverse Drum Clearance. Examine the rivets and the gear teeth on the reverse drum. Try the fit between the drum teeth and the low-speed gear as there should be a clearance of .003 inch on a new job and .005 inch on a repair job. Examine the driven gear to see if it is in good condition and try the keys in the keyways on the gear sleeve and the driven gear. Place and drive the driven gear on the driven-gear sleeve—the outer face on the driven gear should be about .010 inch below the end of the driven-gear sleeve. After assembling, see that all of the gears revolve freely. Assemble the gear shaft to the flywheel; then place the drum assembly, driven gear up, on the bench.

Triple Gear Assembly. Note the punch marks on the triple gears. Assemble the triple gears to the drum gear assembly with the punch marks registering on the triple gears. Assemble the triple gears to the drum-gear assembly, taking note that the punch marks on the triple gears are facing toward the driven gear.

Setting Triple Gears. The setting of the gears may start at any point on the driven gear. The other two triple gears are now spaced by the punch marks, 9 teeth apart, or at 120 degrees from each other. After the triple gears are assembled to the drum assembly, tie a small cord around them so that they will be held in their relative position.

Placing Gear Unit. Pick up the complete gear unit and place the gears down and over the gear shafts and the triple-gear pins. Place the Woodruff keys that hold the disc drum in the gear shaft; the disc drum should then be driven on securely. Place and spread the cotter key so that the set screw does not loosen up and see that the drums are free and that there is not over $\frac{1}{32}$ inch end play in the brake drum. This is very important.

Clutch Assembly. The clutch discs should then be assembled. Place a large disc on first, then a small one until 25 are used, ending with a large one. Replace the push ring and try the fit of the drive plate bushing on the gear shaft; there should be .003 inch clearance on a new job and .007 inch on a repair job.

Fastening Drive Plate. Release the tension on the clutch fingers by compressing the clutch spring and placing the drive-plate cap screw under the clutch shift. The drive plate is then placed in position and fastened down; now remove the temporary drive-plate screw under the clutch shift, also the wire from one drive-plate screw to another. When the clutch is properly adjusted, there should be $\frac{13}{16}$ inch space between the lower side of the clutch and the drive plate. Fig. 35A is a sectional view of the assembled transmission.

Assembly of Transmission. The first parts to be assembled are the reverse drum and gear, driven gear, low-speed drum and gear, and brake drum, as shown in *Group 1*, Fig 35. These parts form *Group 2*. The brake drum should be placed on a bench with the hub extending upward and the low-speed drum should be placed over this hub with the gears on top. The reverse drum is then placed over the low-speed drum with its gear member up. The two Woodruff keys that connect the driven gear to the brake-drum hub are then put in place as shown in *Group 1*. The driven gear is then placed with the teeth downward so that it will be next to the low-speed gear; the triple gears are then meshed with the

driven gear, making sure that the punch marks on the teeth correspond with one another. The reverse gear—the smallest one of
the three comprising the assembly—should be on the bottom, or
down. When the triple gears have been properly meshed, they

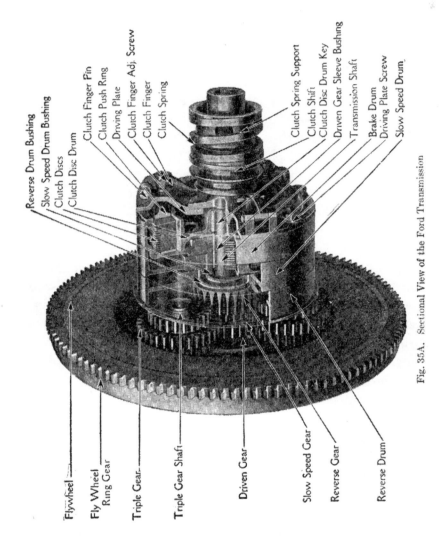

Fig. 35A. Sectional View of the Ford Transmission

should be securely tightened in place with a spring or wire; the
assembly will then be as in *Group 2*.

Group 2 should then be assembled on the flywheel. The flywheel is placed on the bench with its face downward, the transmission shaft projecting upward. The *Group 2* assembly is turned

over so that the triple-gear assembly will face the flywheel. Then the group is so placed on the transmission shaft that the triple-gear pins will pass through the bushings in the triple gears. The assembly will then have the appearance of *Group 3*. The assembly should slip readily in place. If it is necessary to use force, the pins may be bent, or the gears not properly meshed, or the bushings not reamed.

The clutch-drum key should then be fitted in the transmission and the clutch-disc carrier drum placed on the shaft, locking it in place with a set screw provided for that purpose. A heavier disc than the disc plates is put on the clutch drum first; then a small clutch disc, and then a large one. This heavy disc, or distance plate, is not used on the later Ford models. The small and large discs are then added alternately until all the discs are in position. A large disc having keyways on its outer periphery should be on top when the set is assembled. If a small disc having keyways in the inner periphery is left on top, it is likely to drop over the clutch drum.

When changing from high to low speed, it is impossible for the high-speed clutch to be engaged. With the clutch-disc drum and the clutch discs in place, the transmission would have the appearance of *Group 4*. It is then necessary to put the clutch-disc ring over the clutch drum and the clutch push-ring over the clutch drum and on top of the disc ring, with the three pins projecting upward as in *Group 5*. The remaining parts are then assembled in the order shown in *Group 5*. The driving plate should be bolted in position on the brake drum so that the adjusting screws of the clutch fingers will bear against the clutch push-ring pins. It is then advisable to test the transmission by removing the drums and plates with the hand. If properly assembled, the flywheel will revolve freely while any of the drums are being held, or vice versa.

Assembling Clutch. The clutch parts may be assembled on the driving-plate hub by slipping the clutch shifter on the hub so that the small end rests on the ends of the clutch fingers. The clutch spring should then be replaced with the clutch support inside so that the flange of that member will rest on the upper coil of the spring. Next place the clutch spring and the thrust ring with the notch end down on the driving-plate hub and press the spring into place, inserting the pin in the driving-plate hub through the hole on the side of the spring support. One of the best methods of compressing the spring sufficiently to insert this pin is to loosen the clutch-finger tension by backing out the adjusting screw. When these screws are again tightened, the springs should be compressed to a length of 2 or $2\frac{1}{16}$ inches to ensure against clutch slippage. Care should be taken that these screws are uniformly adjusted so that the even compression of the clutch spring is obtained.

Another method is to assemble the spring on the drive plate before installing it on the transmission. A vise or arbor press may then be used to compress the spring. To relieve the pressure on the fingers, a cap screw should be placed between the plate and the shifter.

The clutch has 12 small discs and 13 large discs. A large disc should always be on top when the clutch is assembled.

FORD CONSTRUCTION AND REPAIR

PART II

OVERHAULING FRONT=AXLE SYSTEM

When a repair man starts in to overhaul any part of an automobile, there are usually some definite troubles in a certain part of the car that he is going to look for. In the front axle system, the troubles generally found are as follows: Loose bearings, both in the steering knuckles and in the wheels; wheels out of line with regard to camber, slant and gather; front axle badly bent; and the steering knuckle bushings worn. It is a good policy to examine all parts, even if the axle is only pulled apart for the repair of one part. This will make it certain that there is little possibility of further trouble after the axle is reassembled.

Inspection of Parts. In overhauling the front-axle system, one should first take off the two front wheels and clean the grease from the ball bearings and from the hubs of the wheels. After removing the ball bearings, the steel balls should be carefully inspected for any flaws, pits, or cracks; even a tiny defect is sufficient reason for throwing them away. The surface of a ball is the important part, and a small crack or flaw will cause the cutting of the cone and ruin the entire ball bearing. A cross-section of the front spindle is shown in Fig. 36.

The cones and cups of the front-wheel bearings should be carefully examined for signs of wear. As a rule, it is advisable to replace the old cups and cones with new ones, as worn cones will make it impossible to obtain a satisfactory adjustment of the front-wheel bearings. When the hardened surface of the cone is worn away, the cones will then wear away rapidly.

Turning the cones upside down is sometimes suggested, thus bringing the wear on the opposite side of the bearing. This practice, however, is not to be recommended as it is almost impossible to obtain a satisfactory adjustment, one which will be easy running and not wobble, when the cones are worn to this condition. These cups and cones are shown at *J* and *N* and also at *O* and *E* in Fig. 36.

While the front wheels are off the spindles is a good time to examine and replace the spindle-body bushings and the spindle-arm bushing. In order to remove the spindle-body bushings, it is usually necessary to use one of the special drifts, or punches, which are made for this purpose by some of the accessory manufacturers. If one of these drifts cannot be obtained, the bushings

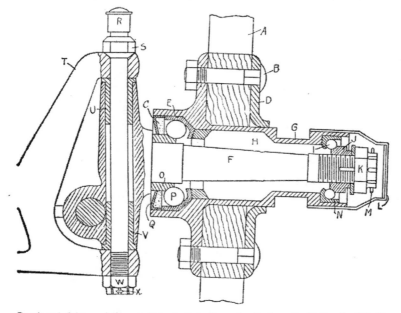

Sectional View of Front Wheel Spindle. A, Spoke; B, Bolt; C, Oil Retaining Wick; D, Hub Flange; E, Outer Ball Race; F, Spindle; G, Hub; H, Grease Space; J, O, Inner Races or Cones; K, Lock Nut; L, Hub Cap; M, Cotter; N, Outer Race; P, Large Ball Bearing; Q, Ball Retaining Ring; R, Spindle Oiler; S, Spindle Bolt; T, Front Axle; U, Spindle Bushing; W, Spindle Bolt Nut; X, Cotter Pin.

Fig. 36.   Cross-Section of Front Wheel Spindle

can be driven out by tapping a $\frac{9}{16}$-inch eighteen-thread tap into a spindle-body bushing and then using an old spindle-body bolt to drive out both the tap and the bushing.

As a rule, it is advisable to replace the spindle-body bolts when the bushings are replaced, especially if the bolts show any sign of wear. After the new bushings have been driven into the spindle body, it is necessary to ream them out, for which purpose special reamers are made and can be purchased at accessory houses.

While working at the front-axle system, the nuts on the end of the front radius rod should be securely tightened and then cotter pinned. If the slots for the cotter pin in the nut do not come into alignment with the holes at the ends of the front radius rod, the nut should be removed and a small amount of metal ground or filed from the face of the nut, after which the nut should be again tightened up. With a little practice, the nut can be adjusted to turn to the correct position by filing off metal in this manner.

The late Fords are being equipped with radius rods fastened to a perch under the axle.

If the nuts on the end of the front radius rods are not kept tight, the vibration on the end of these rods will cause fatigue of the metal and eventual breakage, thus possibly causing an accident. Also, if the nuts on the front ends of the radius rods are not kept tight, the front axle will not be held at the proper slant to give easy and steady steering.

Adjusting Front Axle. *Correct Slant.* The adjustment of the slant of the Ford front axle is of great importance in making the car easy to steer and in saving wear on the tires. If a line is drawn through the axis of the front-spindle bolt, Fig. 37, this line should strike the ground about $1\frac{1}{2}$ inches in front of a vertical line dropped through the center of the axle. By inclining the front axle in this manner, the front wheels are given a trailer, or caster, action, which tends to make them come to a straight-ahead position, just as the front wheel of a bicycle does when the bicycle is held upright and pushed ahead. This steering action of the Ford front-axle system relieves the driver of much strain and fatigue.

Another method of adjusting a front axle to the proper inclination is by tying a weight on the end of a string and using it as a plumb bob. Allow the string to touch the lower side of the front-axle yoke in front of the car, Fig. 38. Under this condition, the string should be about $\frac{3}{8}$ inch, or about the thickness of a lead pencil, from the top arm of the front axle.

Still another way to test this front-axle alignment is by the test method. In using this method, it is necessary to drive the car and see whether the front wheels swing quickly into the

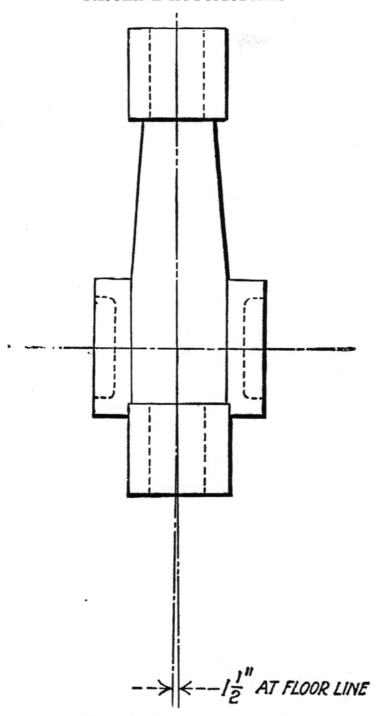

Fig. 37 Checking Adjustment of Front Axle

straight-ahead position after being turned to one side or the other. Of course, this test is made on a smooth level road.

If the front axle is given too much slant, there will be too much tendency for the wheels to come to a straight-ahead position, and the driver will have to exert undue strain and force when turning a corner. For straightaway racing, at high speed on board tracks, it is the custom of some Ford racing drivers to give the front-spindle bolts as much slant as two or three inches, as racing on large tracks does not involve any short or sudden turns.

In order to get the correct slant on the front axle, a large monkey wrench—say a 30-inch size—or a special front-axle tool made for this purpose, Fig. 39, can be gripped at one end of the front axle and used to bend one side of the radius rod, or "wishbone." After bending one side in this manner, the monkey wrench or special tool is shifted to the other side of the front axle, near the other wheel, and then the other side of the axle is given the same slant. The Ford Motor Com-

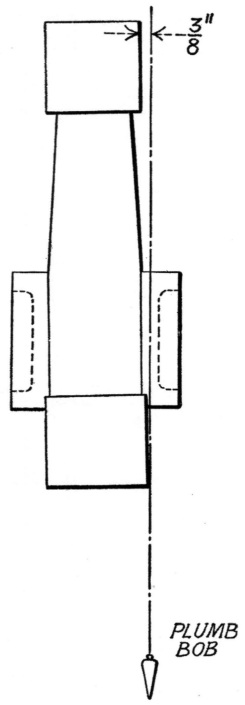

Fig. 38. Checking Front Axle with Plumb Bob

pany recommends that when the radius rods are badly bent they should be replaced with new ones.

Ball Socket. It is usually found that the front-radius-rod ball socket wears loose in time, even though springs are provided under the nuts of the front-radius-rod studs to take up the wear. After a year or so of use, it will be found that these studs are badly worn; as they only cost a few cents each, it is advisable to replace them in order to get smooth steady action in the front radius ball socket.

While the studs are out is a good time to file a little metal from the face of the ball socket on the base of the crankcase. It might be thought that removing metal from the face of the ball-socket cap would be sufficient to eliminate such wear as might be present, still the socket in the crankcase also wears and it is best to file it too.

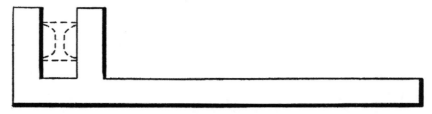

Fig. 39.   Wishbone Straightening Tool

As the bolt holes in the front-radius-rod ball caps are usually badly worn, it is suggested that the ball caps be replaced rather than an attempt made to file the metal off their faces. Of course, if the car has been in use for only a short time, filing some metal from the face of the ball cap will make this joint tight enough for all practical purposes. Much of the rattle is often due to a loose joint in the front-radius-rod ball socket and it is worth while to take particular care in adjusting this socket.

Adjusting Front=Wheel Bearings. After examining the cups of the bearings in the front wheels and replacing such cups as may be worn, we are now ready to replace the wheels on the spindles. If any one of these bearings is not a drive fit into the front hub, several shims may be placed between the bearing cup and the sides of the hub. This is to prevent the bearing cup from turning inside the hub, as all the turning should be done on the ball bearings. The shims used should not be thick or heavy, for if the cup is too tight a drive fit into the hub, the bearing cup may be slightly distorted—out of round—when driven into the hub. It is better to use several thin shims equally spaced around the hub and thus center the ball cup in the middle of the hub than to use a single thick shim at one side. This practice, however, is not recommended.

The adjustment of the front-wheel bearings should be made so that the weight of the valve stem is sufficient to start the wheel in motion and to make the wheel roll to and fro. Yet the bearings should be carefully adjusted so that there is no percepti-

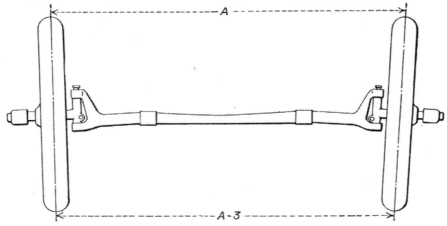

Fig. 40. Adjustment of Front Wheels

ble shake or play when the spokes are gripped and the wheel is shaken.

Before leaving the front wheels they should be checked up for alignment. These wheels should be 3 inches closer together at the bottom than at the top, Fig. 40. The purpose is to bring the point of contact between the tire and the ground more nearly under the point of rotation of the spindle-body bolt. This makes it easier to steer the car and, by having the point of contact between the wheel and the ground come more nearly in the same straight line as the spindle-body bolt, there is less chance of rocks or stones swinging the wheels to one side. This makes it easier to drive the car on uneven rocky roads.

We all know that when a rolling hoop is tilted to one side or the other, it tends to turn a corner or roll to one side in the direction in which it is slanted. The same action takes place in the case of the Ford front wheels, and as the front wheels are slanted outward at the top, they tend to run outward. This tendency would cause undue friction and tire wear if it were not corrected by giving the front wheels a little "gather" to make them $\frac{1}{4}$ inch closer together in front at a point about 16 inches above the ground.

EQUIPPING FORD FRONT HUBS WITH TIMKEN BEARINGS

Sets of Bearings. Ford closed models and 1-ton trucks have for some time past been fitted with special Timken roller bearings on the front wheels. The same bearings can also be installed in other models to replace the old cup-and-cone bearings, as these sets are interchangeable. The Ford Motor Company supplies the bearings in separate packages or cartons; a complete set of bearings for one wheel is in each package. As the Ford spindles have right- and left-hand threads, it is, of course, necessary to supply adjusting cones for the outside bearings with corresponding threads. The packages containing the complete sets of bearings are plainly marked "right wheel" or "left wheel," according to the set each package contains.

Removing Old Bearings. When installing Timken bearings in the Ford front wheels in place of the old bearings, the first step is to remove the old bearings from the wheel hubs and clean out the hub thoroughly so that no grit or metal chips will be left to damage the new bearings. The shoulders of the recesses from which the ball cups were removed should be inspected carefully for high spots, which might cause the cups of the Timken bearings to set high on one side.

The stationary cone is also removed from the inner end of the spindle as it is to be replaced with a special cone. Be careful not to leave rough or high spots on the part of the spindle on which the cone seats, as the Timken inner cone is not pressed onto the spindle, but is a floating slip fit. It has a clearance on the spindle of 0.001 inch.

Installing Cups. Both the inner and the outer cups of the Timken bearings, corresponding to the inner and the outer ball races, or cups, of the old bearings, are press fits in the hub of the wheel. The best way to install them is to draw them both into place at once with a special puller, similar to that shown in Fig. 41. The large or square end of this device is held in a vise or with a wrench while the special handle nut on the other end is turned, forcing the races in position.

Tools of this type can be purchased from the Ford Motor Company, or they can be made in the repair shop from cold rolled steel and a simple forging.

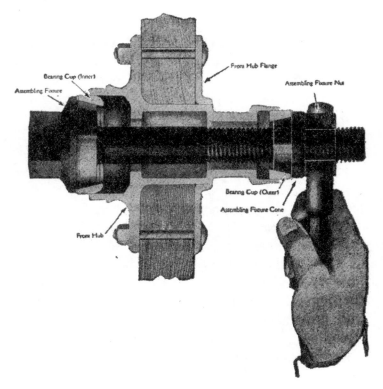

Fig. 41. Tool for Installing Bearing Races

If no special puller is available, the cups can be driven into the hubs, but care must be exercised to drive them evenly all the way around their circumference. A special driver or arbor, Fig. 42, is very useful for this purpose. One end of this driver is used on the large, or inner, cup and the other end on the small, or outer, one. The inside cone faces of the cups must not be struck or marred in any way when pressing the cups into the hub.

Securing Press Fits. As with the cup-and-cone bearings, the Timken cups must be press fits in the hub. It is advisable to try both cups by hand to make sure they will not fit too loosely before attempting to press either one into place. Sometimes a hub will be damaged and the recess for one of the cups expanded somewhat as a result of some of the balls being broken in the old bearing.

If either bearing cup of the Timken set is a loose fit in the hub, it is safest to install a new hub, as there is no satisfactory method of making a bearing cup a tight fit in a hub too large for it. Some repair men attempt to make a cup a tight fit by putting strips of paper or emery cloth between the cup and

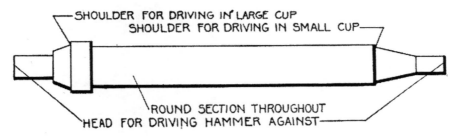

Fig. 42.   Special Bearing Replacer

the hub recess; with either material there is a grave danger of getting the cup out of true. If paper is used, it soon becomes soaked with grease and pounds or works out of place, leaving the cup loose. On the other hand, emery cloth cannot be used at all unless the cup is entirely too loose and sloppy a fit in the hub recess. To put it another way, if the hub is so large that it is possible to use emery paper to make the cup a tight fit, it is not safe to use the hub. The emery is also liable to get into the bearing and hasten the wear.

The depth of the outer race recess in the latest front hubs is $\frac{3}{4}$ inch, but in older hubs it is $\frac{19}{32}$ inch. Some variation may occur, therefore, in the depth to which the outer, or smaller, cups of the Timken bearing sets can be pressed in the front hubs of various cars.

The principal thing to look out for in connection with the outer cups is to see that they are pressed into place evenly and are not high at any spot. It is unimportant if some cups project slightly beyond the end of the hub, are flush with it, or set in slightly, provided they have been pressed in all of the way and run true.

To test the fit of the inner, or large, cup, be sure the distance from the outer face of the cup to the edge of the hub at several different points around the cup is even, Fig. 43. A fine scale should be used, preferably one divided into 64ths, as a very slight difference in the depth of the cup at any one point would cause it to run out of true.

In removing the races used for the ball bearings, the hub surface is sometimes burred. This burr may seem very small and of no consequence, but at the same time it is in a position to cause excessive wear. Care should therefore be taken to prevent these burrs when the old races are being removed.

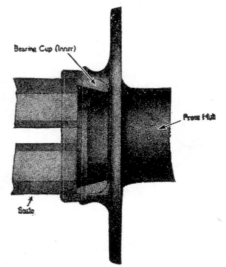

Fig. 43. Checking Evenness of Bearing Cup

If burrs are present they may be removed with a fine chisel and emery cloth. A little sand or grit will cause the same trouble and to avoid the presence of grit, the races and hubs should be thoroughly cleaned with gasoline or kerosene and a stiff brush.

Cones and Rollers. A plentiful supply of a good cup grease should be packed into the hub as well as into the inner and the outer cones and roller sets, the spaces around and between the individual rollers also being filled. The larger inner cone with its rollers is then placed in the inner cup, and the dust ring and the felt washer are driven into the large end of the hub so that the dust cap is flush with the end of the hub.

It will be noticed that the rollers of the Timken bearings are assembled with the cones instead of with the cups, or races, as are the cup-and-cone bearings. For this reason, the large inner cone must be a floating slip fit on the inner end of the spindle body.

The wheel, with the inner bearing complete and the dust ring in place, is next mounted on the spindle. It is never necessary to force the large cone onto the spindle. The outer, or threaded, cone for that side, with its rollers assembled and properly packed with grease, is then screwed onto the outer end of the spindle. A right-hand threaded cone is used on the left spindle and a left-hand threaded cone on the right spindle, as with the old cup-and-cone bearings.

The adjusting cone should be run up on the spindle until the wheel seems to bind slightly. The wheel should then be turned a few times to make sure that all working parts are in good contact, then the adjusting cone should be backed off about one-fourth to one-half turn. This will be sufficient to allow the wheel to revolve freely but without end play.

Sometimes looseness in the spindle-body bushings may be mistaken for end play in the wheel bearings. To avoid any such mistake, insert a cold chisel or a screw driver between the jaw of the axle and the spindle to take up any play that might exist in the spindle bushings, and test the wheel for end play by working the wheel back and forth.

When the proper adjustment has been reached, the spindle washer and the nut should be replaced, the nut being drawn up tight and then cotter keyed as with the cup-and-cone bearings. Make sure that tightening the nut to the proper notch for the cotter key does not cause the bearings to bind; turning the wheel a few times just before the cotter key is inserted will determine this. The hub cap can then be filled with grease and replaced.

Periodic Inspection. Every three or four months the hub bearings should be cleaned out, repacked with fresh grease, and readjusted. The old grease should be thoroughly removed with kerosene or gasoline to make sure that no grit or metal particles remain in the hub to damage the bearings later. The rollers, cones, and cups should be carefully examined for pitting or other signs of wear.

SPRINGS

Replacing Center Bolts. We are now ready to inspect the front and rear springs. The first point is to make sure that the springs are correctly centered in the middle of the chassis frame. Sometimes the center bolts, which hold the leaves of the spring together and keep the spring from slipping from one side to the other, are broken. In such case, the spring leaves or the entire spring is likely to slide a little to one side and cause a slight tilt of the car. These springs are shown in Figs. 2 and 3.

If a center bolt is broken, it should be replaced with a new one. To do this work, it will be necessary for the mechanic to remove the entire spring assembly from the car. While the spring leaves are off the car, the surfaces of the springs should be sandpapered and covered with grease and graphite. The spring leaves will then slide freely over each other, thus making the car easy riding and reducing the likelihood of spring breakage.

Tightening Spring Clips. After replacing the springs on the car, the spring clips which hold the spring in place should be securely tightened, and then cotter pinned. After the car has been in use for some time, the leather pad between the top of the spring and the cross member of the chassis is squeezed down a little thinner. This loosens the springs, and it is therefore advisable to tighten the spring clips again after the car has been run from 500 to 1000 miles. Practically all cases of spring breakage through the middle of the spring are because the nuts on the spring clips have not been kept sufficiently tight. If this is done, the middle of the spring is kept as solid as one piece of metal with the cross member of the chassis and there is practically no bend or flexing at this point. It is then almost impossible to break the spring.

Shackles and Bushings. As the shackles at the ends of the springs come in contact with so much mud and grit, they, and also the bushings in the ends of the springs, are subjected to rapid wear. This wear will go on at an increasing rate if the worn parts are not replaced. It is a comparatively easy matter to drive out the old bushings from the ends of the springs and to drive in new ones; and about the only advisable repair on the spring shackles is to replace them with new ones. A very handy method of replacing the spring bushings is shown in Fig. 44. A tube larger and longer than the bushing is placed on one side of the spring opposite the bushing to be removed. The new bushing is placed against the one to come out and the three parts are then caught between the jaws of a vise. As the jaws are screwed together, the new bushing forces the old one out of the spring; then the new bushing is reamed to size. A reamer is not expensive and should be included in the Ford mechanic's tool equipment to secure the best results. The late Fords are equipped with steel bushings in the springs, which require no reaming.

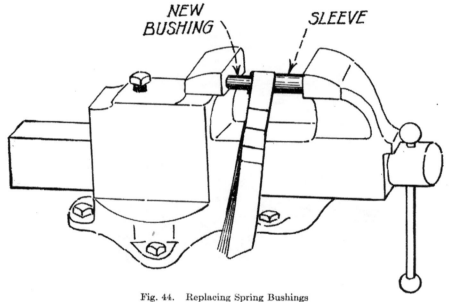

Fig. 44.   Replacing Spring Bushings

OVERHAULING REAR=AXLE ASSEMBLY

Removing Rear=Axle Assembly. With the car jacked up at the frame, the rear-axle assembly is taken out as a unit. The two bolts and the cap screws holding the universal joint should be removed, the brake rods disconnected at the front end, and the spring shackles removed. The nuts on the front end of the radius rods near the universal joints, Fig. 45, are then removed, and after the nuts, Fig. 46, at the rear of the drive shaft holding the drive shaft in place are removed, the drive-shaft assembly can be moved toward the front. The wheels are then taken off. The differential housing is held together by seven bolts, and, after removing the axle assembly from the car, these seven bolts should be removed. Put the nuts on the bolts and keep them together so that they can be found when needed again—in fact, this should be done with all the removed bolts and nuts. Whenever it is practical to put the nuts and bolts back in place after disassembling, it should be done, as much time will be saved. When the seven bolts have been removed, the two main parts of the housing

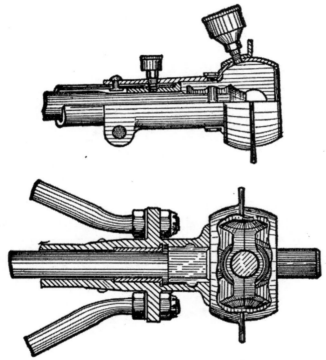

Fig. 45. Sectional View of Universal Joint

can be separated, disclosing the differential assembly and the bevel-gear driving system.

Bearings. The inner and the outer shaft bearings are in the housing. The sleeves forming the outer parts of these roller bearings are forced tightly into place, and if an attempt is made to remove them, they will be spoiled. They are split and are inserted at the factory with the split edges lapped; these sleeves may be installed with the use of a hammer. In removing these sleeves, they will be bent or sprung so much that they cannot be restored to their original accuracy if a special puller made for this purpose is not used. If there is wear, the spoiling of the sleeves does not matter as it will be necessary to install new ones. There is no way to compensate for the wear in these bearings.

The rollers, however, come out easily. Because of the ample length and the extreme hardness of the rollers, they are likely to show less wear than the sleeves or the live axle shafts. The rollers bear directly on the live shafts, and as the shafts are necessarily rather soft, there will be no tendency toward brittleness.

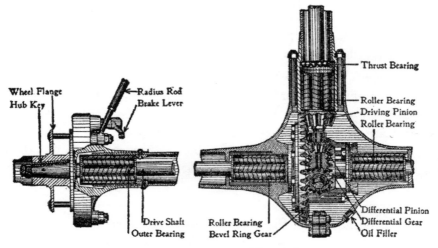

Fig. 46. Sectional View of Rear Axle

The shafts may therefore be expected to show more wear than the other bearing parts. This naturally means new shafts.

The rear-axle bearings should then be inspected. If there is play in the parts of the inner roller bearings, the axle shafts may not stay exactly in the center of the axle housings. The amount of play should then be adjusted so that it is the same in one direction as in the other.

After getting the axle shafts centered in the middle of the axle housings, the rear nuts on the radius rods should be adjusted against the shoulders of the drive-shaft housing as shown in Fig. 45 near the universal joint. After having this adjustment correctly made, the nuts on the front end of the radius rod should then be tightened up securely. If these nuts are not kept very tight, the radius rods will rattle and pound at this point. The hammering of metal against metal will also tend to cause crystallization and fatigue of the steel—a frequent cause of the breakage of Ford radius rods.

There is little use in half doing a replacement job of this kind. If there is looseness in the bearings and the sleeves and shafts show wear, both should be replaced. If only the sleeves are replaced, there will still be looseness caused by the wear of the shafts, which will quickly develop into more looseness. In short, it will not pay. While at it, put in whatever new parts are needed and make the job right. If this is done, there will be practically a new outfit.

Wear of Thrust Rings. The thrust of the bevel driving gears creates a tendency for the large driving gears and the drive pinion to move apart. In other words, there is end thrust. This thrust operates against the thrust rings placed between the differential case and the rear-axle housing. The rings are of bronze or babbitt and are grooved to distribute the lubricant. The wear on the rings varies considerably, depending on the kind of roads on which the car is run, its load, the way it is driven, etc. If there is wear, the rings should be replaced with new ones.

This ring wear allows the bevel driving gear and its pinion to separate, consequently the teeth do not mesh as deeply as they should and there is a reduction of tooth surface in contact, wear of the teeth, inefficient operation, and noise. The bronze or the babbitt rings are placed between steel rings pinned to the differential cage and to the housing. Thus there are six rings: two floating rings, two rings on the differential cage (one on each side), and two rings on opposite sides of the axle housing.

Differential Gears. The differential gears are in a cage which carries the large bevel driving gear. The cage is made in two halves, and it can be taken apart readily. It is held by three studs and nuts, and once the nuts are removed, the spider carrying the three small differential pinions is released, as the spider is held in place by the two castings that form the differential cage. This assembly is shown in Fig. 17. The pinion bearings should be examined for wear as more or less wear is likely to be found. The pinions themselves are of hardened steel and as a rule do not wear much; ordinarily they are found in good condition. Look at the bearings, however, and replace them if worn.

The live shafts and the large gears of the differential can be removed from the differential cage when the cage has been separated, which is done by pulling the shafts through the inside of the cage halves. The gears can be removed from the shafts after the shafts have been taken out by first forcing the gears farther on the shafts, permitting the removal of the split rings, and then sliding the gears off and leaving the keys in their keyway.

Ring Gear. The large driving, or master, gear is bolted to the differential spider, so that it can be replaced when it becomes worn. A new gear should be used if the teeth of the old one are chipped, burred, or worn.

Pinion Gear. The bevel driving pinion, Fig. 47, on the rear end of the propeller shaft is subject to greater wear than the ring gear because it has a smaller number of teeth and therefore a smaller surface over which the wear is distributed. So it is reasonable to look for some wear in the pinion even if the ring gear is in good condition. Do not use a pinion that shows signs of wear or is chipped, as it deteriorates quickly when once it starts to chip. Removing the pinion is accomplished by taking out the locking cotter pin that holds the castellated nut, after which the gear may be taken off with a gear puller.

Rear Axle Gears. The bevel gear in the rear axle has 40 teeth and the pinion has 11 teeth. The gear ratio is $3\frac{7}{11}$ to 1. When the car is being operated in low speed, the gear ratio between the motor and the rear wheels is 10 to 1; and when reverse speed is used, the ratio is 14.5 to 1.

Removing Drive Shaft from Housing. To remove the drive shaft from the housing, it is first necessary to remove the universal joint from the front end of the shaft. This universal joint is held on the shaft by a pin. The shaft end and the socket are square thus preventing turning.

To remove the pin, two plugs—one of which is located at the top and the other at the bottom of the shaft housing, near the universal joint—must first be removed. The bottom plug is shown in Fig. 48. The pin can then be driven out with a punch, after which the universal joint will come off when tapped slightly with a hammer. The drive shaft can then be removed at the gear end of the shaft housing.

Drive=Pinion Bearings. Perhaps the most important bearings in the rear-axle assembly are those back of the drive pinion. There is a ball-thrust bearing and a roller bearing, and these should be thoroughly inspected to see that there is no lost motion; if there is, put in new bearings as the alignment of the gears depends on the fit here. Even if everything else is in perfect condition, looseness of these bearings will cause grinding and rapid wear when running.

Fig. 47. Driving Pinion and Ring Gear

Caution. If any of the gears have been broken or the teeth chipped, some of the pieces may have lodged between other teeth, and they will do a great deal of damage in this position.

PIN PLUG HOLE

Fig. 48. Universal Joint and Housing

The gears should therefore be carefully inspected and any chips removed.

Models Differ. In the late bearing models the seats in the central section of the axle housing carrying the inner bearings are of pressed steel fitted into the castings and held in place by the same rivets that hold the tubes to the housing castings. In the early models these seats are formed directly in the castings by machining. This makes no difference in the disassembling or the assembling of the rear axle, but it is a point that is well to mention to avoid any confusion.

Adjustments. In overhauling a Ford rear axle, it should be borne in mind that while no means of adjusting the mesh of the gears is provided, all parts subject to wear are removable and renewable. The non-replaceable parts are not subject to wear. Therefore, if a rear axle is fitted with a new set of wearing parts, it will be in as good running condition as when it left the factory, providing the rear-axle housing has not been sprung or deformed by accident. All the parts are so easily obtained and installed that there is no reason for not making the necessary renewals. Saving money by not installing or renewing worn parts is very expensive economy, for it will lead not only to excessive wear of the parts in question but to the imposition of extra strains on other parts, causing them to wear more rapidly than they should.

For example, consider the thrust collars. If these collars are loose and other parts fit properly, there will be end play. This end play will allow the gears to work away from each other, placing extra wear on the teeth and increasing the thrust on the already worn collars and on the important bearings back of the driving pinion on the rear end of the propeller shaft.

Referring again to the bevel pinion, do not try to remove it by driving unless the propeller shaft is removed and stripped; trouble may follow. Use a wood block to drive the shaft out of the pinion so that the shaft end will not be damaged. The pinion is mounted on a taper shaft, and once it is started, it drops off. Do not use the Woodruff key—which will prevent the pinion from turning on the shaft—until it is examined to make sure that it is not battered or worn so that the pinion can move on the shaft. Be very sure that the key is in place when the pinion is put back; if not, there can be but one result—the car will not run.

Clean every individual part thoroughly and scrupulously. Do not leave a trace of old oil or dirt anywhere. When putting the parts together again, oil them, so there will be no danger of rusting the surface of the parts that are out of reach of the oil in the housing. See that the spiral rollers of the roller bearings are thoroughly cleaned inside.

Reassembling. In reassembling the axle, make sure that all parts go together as they originally belonged. If anything binds or sticks or will not go in as it should, there is a reason. Find the reason and remove it instead of trying to use brute force. Make sure that every nut, every screw, and every bolt is properly tightened. At the same time, do not make the mistake of tightening nuts with so much force that the threads are partly stripped. Be sure that all cotter pins are replaced and that no passages designed for lubrication are blocked. Finally, see that this system is given a plentiful supply of the right lubricants. The Ford rear axle needs plenty of lubrication, and the owner should see that it gets it.

LUBRICATION SYSTEM

Importance of Lubrication. Lubrication was the subject of a very thorough investigation when the Ford unit power plant was designed because, while the main purpose was to obtain extreme manufacturing and operating simplicity, sufficient lubrication at all times had to be provided. From the viewpoint of an engineer, excessive lubrication of an internal-combustion engine will not have destructive influence, although accumulations of oil in the cylinders will coat the spark plugs and will impair, if not totally prevent, ignition. The unconsumed lubricant will collect upon the piston heads and the combustion chamber and will, with dust and foreign matters drawn through the carburetor air intake, become burnt and hardened by the heat and form carbon. While such a condition will lessen the efficiency of the motor, no actual damage will result unless the motor misfires in which event extra strains will be placed on the rear axle and the engine.

But if the oil supplied is insufficient, damage will certainly result, the most probable effect being the heating and scoring or even the melting of the babbitt-metal bearings of the crankshaft and the big ends of the connecting rods. Far more serious consequences may happen, such as a piston seizing in a cylinder, which may possibly buckle the crankshaft, bend or break a connecting rod, break a piston, or even puncture a hole in the crankcase, in the event that a connecting rod is broken.

When a bearing is not properly lubricated and it heats until the babbitt metal is softened or flows from the cage to the rod end retaining it, it is referred to as a *burnt-out* bearing. In other words, because of lack of oil, the friction so heats the metal that it becomes plastic and no longer supports the load upon it, taking a new shape and so enlarging that the shaft or rod has side or end play, which causes a noise of a peculiar and noticeable character. Obviously the only remedy in such a case is the replacement of the bearings—not a matter of great cost for parts, but quite an expense for labor and loss of service until the replacement is made. The standard labor price for this operation is $4.50.

Continuous Lubrication. Attention has been directed to the consequences of excessive and inefficient lubrication as adequate lubricity is imperative, and this cannot be obtained unless decided care is taken, despite the simplicity of the Ford system. In theory and in practice, the best results can always be obtained with machinery by feeding the lubricant in small quantities, constantly, and by having a supply which can be drawn upon for a considerable period of time. In motor-vehicle practice, the last-mentioned requirement is very important, for frequent renewals would demand an amount of care that would be objectionable to the drivers. The purpose of the design of the Ford engine was to have a system which would feed oil continuously and which could be operated for a considerable time or distance with one supply, the replenishments depending largely upon the use made of the car.

Simplicity and economy demand that the system have the fewest parts practicable, and efficiency requires that the engine be fully lubricated, especially as many of the owners have little or no knowledge of mechanics and might, because of ignorance, neglect conditions that would receive attention from those more experienced.

Parts of Lubrication System. A sectional view of the Ford motor, Fig. 49, shows the lubrication system. As previously stated, the engine block and the head are cast separately, the cylinder block forming the upper portion of the crankcase, while the lower half of the crankcase is of pressed steel, about $\frac{1}{16}$ inch thick, extended back of the engine to form the bottom section of the flywheel and gear-set case. In this pressed-steel section is an opening extending from a point just forward of the front wall of No. 1 cylinder to a point directly beneath the wall between cylinders 3 and 4. This opening is $13\frac{5}{8}$ inches long and $5\frac{1}{4}$ inches wide and is surrounded by a raised edge, and a ring that is $\frac{3}{8}$ inch high and $\frac{5}{8}$ inch wide.

Oil Troughs. The opening is closed by a pressed-steel plate, or cover, $15\frac{1}{3}$ inches long and $6\frac{7}{8}$ inches wide, bolted on with a gasket between it and the case, thus making an oil-tight joint, the lap being about $\frac{3}{4}$ inch. The plate is slightly curved to conform with the general shape of the crankcase, and in the plate are three transverse troughs $\frac{9}{16}$ inch deep that are, when the cover is in place, directly under the caps of the connecting rods of the first three pistons of the engine. These can be noted in Fig. 49.

Oil Reservoir. From a point just back of the rear portion of the ring about the opening, the crankcase is sharply bellied, or enlarged, to form a housing for the flywheel, and directly under the flywheel there is a cone-shaped pocket. From this the crankcase bottom rises to a point slightly above the level of the connecting-rod caps. From the ring rearward the bell housing of the flywheel forms the reservoir in which the oil is carried.

Correct Level of Oil. There are two drain cocks located in the rear of this oil reservoir, so that the amount of oil contained in the tank can be ascertained. The motor must never be run until the oil is below the level of the bottom cock. When starting on a trip, the reservoir should be filled so that the oil will run out of the top cock. Be sure that dirt has not obstructed the openings in these cocks, as they are subjected to a great deal of road dirt. To secure the best results, the oil level should be about halfway between the two gage cocks. Were there exact knowledge of the quantity of oil required to fill the crankcase between the two drain cocks and were half of this oil placed in the engine base, perhaps the proper level would be reached, but there is no way of determining the consumption of the lubricant other than to learn if there is a flow from the lower cock.

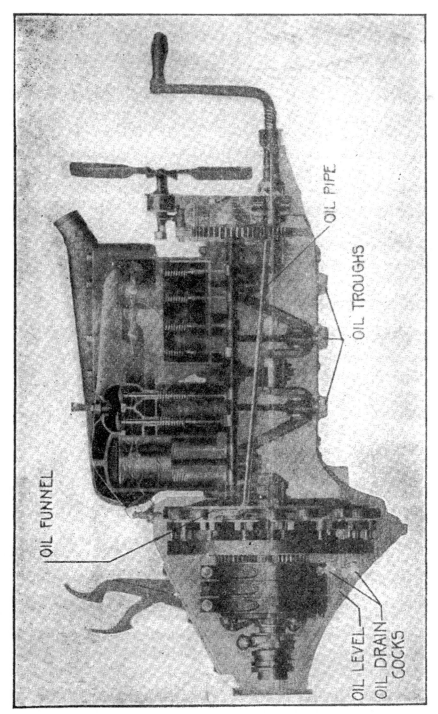

Fig. 49. Power Plant Lubrication System

It should be remembered that the crankcase of the engine is not obstructed below the level of the main bearings from end to end and that the flywheel edge as it revolves is about $\frac{1}{2}$ inch above the inclined bottom of the case behind it. The engine case when full will contain about 4 quarts of oil, the lubricant being $\frac{15}{16}$ inch deep in the troughs behind the three front connecting rods in the bottom cover plate and slightly below the level of the cover plate. The troughs are supposed to contain sufficient oil to permit the connecting rods to dip into it as they revolve. When the machine is ascending or descending grades, the flow of the lubricant in the troughs must be either backward or forward, and in volume depending upon the angle of the grade.

Circulation of Oil. Assuming that the machine is being driven on a level surface, the condition of the oil in the crankcase

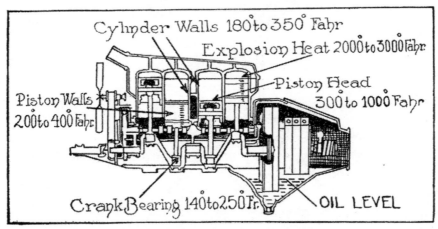

Fig. 50. Motor Temperature and Lubrication

is similar to that shown in Fig. 50. About $\frac{1}{3}$ of the diameter of the flywheel is submerged in the oil; and the lower part of the three contracting bands that encircle the revolving planetary-gear set, the peripheries of the high- and the low-speed drums, and the surface brake drum are sprayed with oil from the flywheel. At every revolution of the crankshaft, the caps of the three forward connecting rods strike the lubricant in the troughs, while the cap of the rear connecting rod strikes a heavy spray of oil in the reservoir ahead of the flywheel. The approximate temperature of the various motor parts is also shown in Fig. 50.

As the engine turns, the sweep of the connecting rods into the oil creates a splash that throws the lubricant from the left to the right side of the crankcase, and as the ends of the rods are swung through the space beneath the cylinders, the greater part of the oil is thrown off in the form of a spray. This is the result anticipated in all splash systems. The revolution of the gear set does nothing more than plentifully lubricate the pinions and the gears, and the degree of the lubrication is greatly in excess of the actual needs, but this is not a condition that can be criticized as it is insurance against wear.

As the flywheel revolves in the oil, it carries upward considerable lubricant, which is thrown off against the right side of the crankcase and the top of the flywheel housing. A small funnel is attached to the inside of the crankcase, Fig. 49. This funnel is directly in the line of the movement of the flywheel assembly, and the legs of the sixteen magneto magnets, clamped to the flywheel, serve as paddles and throw the oil up in considerable quantity when they rise above the surface of the lubricant; part of this flows into the funnel. The funnel is connected with a brass tube that leads along the inside of the crankcase and all the oil collected is carried forward to the gears at the front of the motor in the timing-gear case. When the crankcase is filled to the level of the highest drain cock, the volume of the oil circulated will be the greatest, and with the engine running at 1500 r.p.m., the circulation will be at the rate of about 2 quarts per minute. As the oil is consumed and the level is lowered, the volume of the oil circulated will decrease—probably to less than half the maximum of 2 quarts, and when below the lower drain cock, the circulation lessens rapidly. Of course, there are other conditions that influence the oil circulation, the character of the oil—for a heavier lubricant will not be as thoroughly distributed, and a lighter oil will be carried in a larger volume—the heat of the engine, temperature of the air, all are factors that must be considered.

Filling the Troughs. As the oil is carried forward in the tube, it floods the timing gears and then flows into the bottom of the crankcase and over the forward end pan, filling the pool beneath the connecting rods. The connecting rods dip into the oil about $\frac{1}{2}$ inch when the piston is on lower dead center. In recently built chassis, openings are made in the sides of the oil troughs, varying from $\frac{1}{16}$ to $\frac{1}{8}$ inch, to reduce the depth of the oil beneath the connecting rods, but obviously the flow is greatly dependent upon the heat of the engine, the grade and viscosity of the oil, and the volume supplied. As the openings are not uniform in width, one cannot determine what will be the actual oil depth in the troughs in any given operating condition. The oil thrown off by the big ends of the connecting rods lubricates the center and the rear main bearings, the wristpins, cylinders, pistons, cams, and valve tappets.

Viscosity of Oils. The oil used for lubricating purposes is intended to form a film between two moving surfaces, and by preventing actual contact of the parts it minimizes friction. The fluidity of oil is spoken of as its viscosity and is measured by the time required for a given volume at a given pressure to pass through a standard aperture. The time is expressed in seconds and the reading is usually taken at 200 to 212°F. The range of the test of oils used in internal-combustion motors is from 180 seconds for a light or medium oil to 2300 seconds for an extremely heavy oil. Oils of less than 180 have insufficient body to lubricate satisfactorily and those of more than 800 are unsatisfactory because the fuel consumption is increased. It has been found by laboratory tests that maximum results can be obtained in the cylinder with oil having a viscosity of 180. This affords the greatest horsepower for the amount of fuel consumed.

Formation of Carbon. Oil from the lubrication of the pistons and the cylinders is splashed on the lower cylinder walls and is carried upward and spread over the cylinder walls to the height of the piston stroke by the pistons and the piston rings. A certain quantity is thrown off the pistons by the upward strokes and is projected onto the walls of the combustion chambers. If the quantity thrown off is small and the mixture is "lean" and is consumed rapidly, the oil will be practically all burned by the explosion and there will be no appreciable deposit of carbon. But when the quantity is large and the heat of the explosion does not consume it as readily, the vaporized portion is exhausted with the gases as smoke and the remainder is left on the walls as the heavy end-products of destructive distillation. These are reduced by the intense heat into a cumulative incrustation that is generally referred to as carbon deposits.

Effects of Carbon. The deposits of carbon in the combustion chamber, on the valves and seats, the spark plugs, and the piston heads decrease the efficiency of the motor; and while burning a mixture of fuel that will as far as possible secure complete combustion and thorough scavenging of the cylinders will undoubtedly have some influence, carbonization will eventually result. Yet the use of a good oil will greatly lengthen the period of service between removals of carbon. Sooted spark plugs, necessitating frequent cleaning and causing faulty ignition and loss of power; carbonized valves and valve ports, followed by leakage, loss of compression, dilution of fuel, and excessive fuel consumption; preignition from the points of carbon becoming incandescent—all these are among the certain results of carbonization.

Cylinder Oil Film. The cylinders of the engine are usually bored to have the same diameter the entire length. The pistons are generally turned to have a slightly smaller diameter at the top, or head, to allow for expansion. The cylinder walls will be kept reasonably cool by the circulation of the water, but the pistons are cooled only by the admission of cool fuel and by the splash of oil into the cylinders. The clearance—the space allowed between the walls of the pistons and the cylinder—is filled by the piston rings, which are formed to be slightly larger than the cylinders and are compressed into the ring grooves. If the cylinders are true and the piston rings fit perfectly, the latter will prevent the escape of the gas during the compression and the explosion strokes, the film of oil between the rings and the cylinder walls forming an oil seal. The lubricant that best serves is that which affords the most perfect seal and the greatest degree of lubricity.

CARBURETION SYSTEM

Importance of Correct Mixture. Many mechanics do not realize the importance of a perfect carburetor adjustment. Different adjustments should be used when the car is driven under certain conditions. For instance, a certain amount of gasoline can be saved if the mixture is thinned down when the car is being driven between towns and a regular speed can be maintained. The pick-up will not be quite so good, but this sacrifice can be made in view of the saving in gasoline.

The experienced driver can tell from the sound of the motor whether it is laboring with an over-rich gas mixture or whether it is "starving" for want of a mixture sufficiently rich to give the motor full power and to obtain gasoline efficiency.

Troubles Misleading. Many ignition troubles have symptoms similar to those of carburetor troubles, and it sometimes takes a little time to determine which is at fault, the carburetor or the ignition system. For example, a poorly adjusted carburetor and a weak magneto have similar symptoms, as a weak magneto will not satisfactorily fire a charge that is either too rich or too lean.

If the car is taken out on the road and run at about 15 m.p.h., the carburetor can be adjusted so closely that the car will run perfectly at that speed. As soon as the car is stopped and again started, the same trouble may recur, and the motor may misfire a great deal.

Back Fire in Intake Manifold. Too thin a mixture will make the motor spit-back or back-fire in the intake manifold and into the carburetor. The layman and even the mechanic will often wonder how the gas in the intake manifold and in the carburetor can be burned as the explosion takes place in the cylinder long after the intake valve has closed. It must be remembered that a thin mixture burns much slower than either the proper mixture or the rich one.

There is a lapse in time of only a small fraction of a second between one explosion and the opening of the intake valve at the beginning of the next cycle when the motor is running at 1200 revolutions per minute, or 20 revolutions per second. As there is one explosion in each cylinder in every two revolutions, 10 explosions will occur in that cylinder in 1 second; in other words, $\frac{1}{10}$ second will be the time allowed for the completion of the four strokes of the cycle. This allows $\frac{1}{40}$ second for each stroke of the piston. When the spark occurs, the thin mixture continues to burn through the power and the exhaust strokes, and it is still aflame when the intake valve opens at the beginning of the next cycle. This gas has then been burning through two strokes of the cycle, or $\frac{1}{20}$ second, if the motor is running at 1200 revolutions per minute. The gas in the intake manifold is ignited by this flame.

Carburetor Adjustments. There are but two parts that can be adjusted on the Ford carburetor, the float, which needs very little adjustment, and the needle valve. In order to adjust the float, it is necessary to take the carburetor apart. The float,

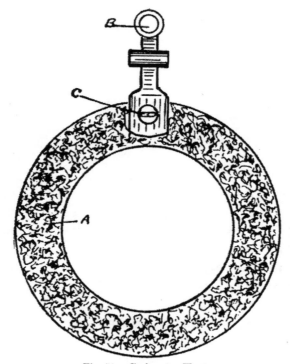

Fig. 51. Carburetor Float

Fig. 51, is made of cork and is well shellacked so that the gasoline will not be absorbed and cause the float to be heavy, or waterlogged. The float closes the valve that allows the gasoline to enter the carburetor from the gasoline tank, and if the float is waterlogged, the valve will not close when the gasoline has reached its proper level. This will cause the carburetor to leak when the car is standing and permit a rich mixture when the motor is running. To remedy this trouble, the float must be thoroughly dried and a fresh coat of shellac applied; or a new float may be installed. The float arm must be bent if the gasoline level in the float is not at the proper point. Fig. 52 is a sectional view of the Kingston carburetor, while the Holley carburetor is shown in Fig. 53. The gasoline level must be high enough to allow the gasoline to come just below the top end of the spray nozzle. If this level is low, the motor will start hard. If the float valve C, Fig. 52, does not seat properly, or leaks from any cause, the carburetor will leak just as though the float was water-logged. The pin B holds the float and the float lever in proper relation to the float valve C. The gasoline, or needle-valve,

adjustment is made by turning the needle J, the adjustment rod extending to the dash of the car. If the needle is turned to the right, or clockwise, the mixture will be made thinner; if turned to the left, or anti-clockwise, a richer mixture will result.

Care of Gasoline Line. The gasoline line is a very important part of the carburetion system. It is sometimes clogged or

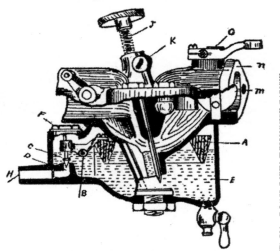

Fig. 52. Sectional View of Kingston Carburetor

obstructed with foreign material, thus preventing the gasoline from flowing properly to the carburetor. Lack of gasoline will cause the motor to spit-back in the intake manifold just as it would if the nozzle was set for too thin a mixture. If the gasoline line is obstructed, the motor will generally run without missing at low speeds, but when speeded up, it will start to miss. This is due to the fact that there is not enough gasoline flowing to the carburetor to replace the gasoline that the motor has consumed.

Care of Spray Nozzle. The carburetor spray nozzle has a very small opening and therefore a particle of some foreign substance can easily clog up this opening. This will cause the motor to misfire and slow up when the throttle is opened, the motor acting much as if there was a clogged-up gasoline line. This occurs because the motor requires a great deal more gasoline when running fast than when idling. The obstruction can generally be removed by opening the needle valve a few turns and then opening the throttle several times in rapid succession; this draws the dirt or other foreign matter through the spray nozzle. The needle must then be turned back to the original position. If this does not remove the obstruction, it will be necessary to drain

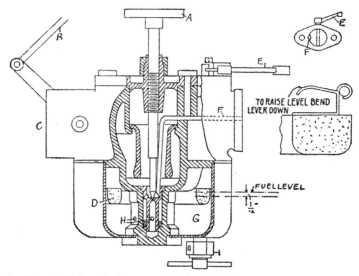

Section of Holley Carburetor: A, Needle Valve; B, Choker Rod; C, Auxiliary Air Intake; D, Float; E, Throttle Lever; F, Slow Speed Supply Tube; G, Float Chamber; H, Supply Holes to Needle Valve; I, Drain.

Fig. 53. Sectional View of Holley Carburetor

the carburetor, which, after it is removed from the car, should be thoroughly cleaned.

Hot=Air Pipe. The hot-air pipe furnishes a supply of warm air to the carburetor and vaporizes the fuel. It is advisable to remove the hot-air tube during the summer months, but this tube must be on the car during the winter months.

Throttle Adjustment. If the motor runs too fast when the throttle is fully closed, it is necessary to adjust the throttle stop screw. The lock screw that keeps the throttle-adjusting screw from turning, Fig. 52, should first be loosened. The throttle screw is then turned out—anti-clockwise—until the motor slows up to the desired speed. The lock screw is then tightened so that the adjustment will be permanent.

Setting Carburetor for Heavy Fuels. The old Holley carburetors were fitted with a strangling tube $\frac{13}{16}$ inch in diameter at the throat. This strangler, or mixing tube, was satisfactory for the high-grade fuels of a few years ago, but it does not handle the present fuel as it should. This mixing tube, Fig. 54, can be replaced with a tube $\frac{23}{32}$ inch in diameter at the throat for the proper mixing of the present heavy fuels.

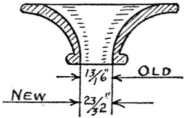

Fig. 54. Carburetor Mixing Tube

The smaller tube causes a greater velocity of the gases through the mixing chamber, and therefore there is a better mixture when the gas enters the cylinders, which results in greater mileage and power.

OPERATION OF FORD CAR

Introduction. Operation does not mean the mere pressing of pedals and pulling of levers; it means also the why and wherefore of these actions and a knowledge of what is taking place inside the motor, the transmission, the rear axle, etc., when the various pedals and levers are moved.

Preliminary Inspections. *Cooling System.* Before trying to start the engine, the radiator should be examined to see that it is full of clean water. If perfectly clean water is not obtainable, it is advisable to strain the available water through muslin or other similar material so that foreign matter will not get into the circulating system and obstruct the small passages in the radiator.

The cooling system holds approximately 3 gallons of liquid. It is very important that the cooling system be filled before the motor is operated as the motor will become too hot if this is not done. When the cooling system is completely filled, water will run out of the overflow pipe.

The motor will naturally use more water during the first few days of its operation—because it is a new motor—as during this time the parts are fitting themselves to each other and more heat is naturally developed. If it is possible to secure rain water, by all means do so, for rain water does not contain alkalies or minerals which tend to deposit sediment and start corrosion in the cooling system. A phantom view of the cooling system is shown in Fig. 55.

Gasoline Supply. The gasoline tank should next be examined to see that there is a sufficient supply of fuel. In filling the tank, the gasoline should always be strained, preferably through chamois skin, as this prevents water and other substances from getting into the tank. Dirt or water in the gasoline is sure to cause

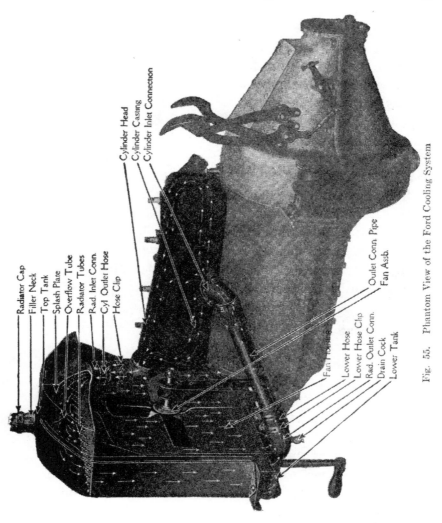

Fig. 55. Phantom View of the Ford Cooling System

trouble. The small vent hole in the gasoline-tank cap should not be allowed to get plugged up as this would prevent the proper flow of the gasoline to the carburetor. There is a drain cock at the bottom of the tank which may be opened to allow the removal of sediment or foreign substances which collect in a dirt trap above the drain cock. The capacity of the gasoline tank when full is about 10 gallons. A sectional view of the fuel system is shown in Fig. 56.

Oil Supply. The supply of oil should next be inspected. There are two pet cocks underneath the car in front of the flywheel housing which are used as gages to show the supply of oil in the reservoir. The upper cock should be opened and a medium grade of good oil poured into the breather pipe until the oil flows out of this cock. After the engine has been limbered up, best results will be obtained by carrying the oil level midway between the two cocks, but it should be borne in mind at all times that, under no circumstances, should the oil be allowed to get below the lower cock.

Control Levers. The next move is to examine the control levers, which are located underneath the steering wheel.

The right-hand lever controls the throttle and is used to regulate the amount of gas vapor allowed to pass into the cylinder. When the engine is in operation, the farther this lever is moved down, the faster the engine will run and the greater will be the power developed. The throttle lever should be opened about five or six notches. This position varies according to the setting of the carburetor and the condition of the motor; a little experience will rightly determine the best position for each individual car.

The left-hand lever controls the time that the spark occurs in the cylinder in relation to the position of the piston. By pulling this lever down, the spark is advanced and by moving the lever up, the spark is retarded. In setting the levers before starting the motor, the spark should be put in about the third or fourth notch of the quadrant.

The hand lever which comes through the floor boards at the driver's left should be inspected to see that it is pulled all the way back as it holds out the clutch when in this position, thereby disconnecting the motor from the rear axle; at the same time it applies the emergency brake at the rear wheels.

Starting the Motor. *Car with Starter.* The ignition-switch key should then be inserted in the switch and turned to the battery position (if the car is equipped with a starting and light-

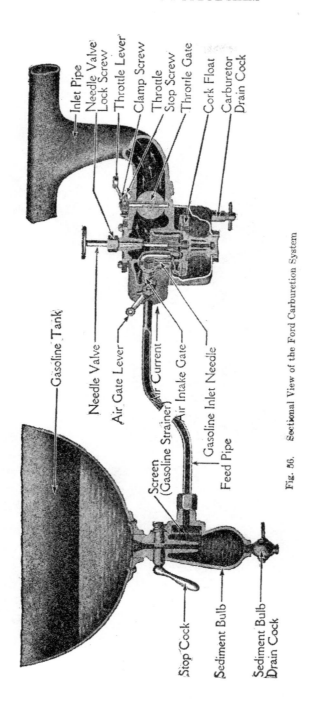

Inlet Pipe
Needle Valve Lock Screw
Throttle Lever
Clamp Screw
Throttle Stop Screw
Throttle Gate
Cork Float
Carburetor Drain Cock

Gasoline Tank
Needle Valve
Air Gate Lever
Air Current
Air Intake Gate
Gasoline Inlet Needle
Screen (Gasoline Strainer)
Feed Pipe
Stop Cock
Sediment Bulb
Sediment Bulb Drain Cock

Fig. 56.   Sectional View of the Ford Carburetion System

ing system at the factory). The choker should be pulled out if the motor is cold and held in this position while the starter button is depressed.

Fig. 57.   Dash View of Cars not Equipped at the Factory with a Self-Starter

The storage battery then furnishes energy to the starter, thereby cranking the motor at a rate of speed sufficient to start its operation. If the motor does not start at once, do not hold out the choker, as this will flood the combustion chambers with too rich a mixture; if the motor starts and then stops, this is an indication that the motor has either been starved or flooded. A cloud of heavy black smoke having a gassy smell will come out of the muffler if the motor has been given too rich a mixture.

Car without Starter. If the car is not equipped with an electric starter, it will be necessary to hold out the choke rod which projects through the front of the radiator while the motor is being cranked. It is also advisable to open the needle valve, say $\frac{1}{4}$ turn, until the motor is warmed up. This is especially

Fig. 58.   Dash View of Cars Equipped with Self-Starter

true during cold weather. If the motor as a rule is very hard to start during the winter months, a good method to use is to crank the motor several times with the ignition switch off, holding out the choke rod while so doing. Then turn on the ignition switch to the **MAG** position and crank the motor. When stopping the motor, it is common practice, especially in cold weather, to pull out the choke at the front of the radiator instead of turning off the ignition switch. This operation leaves a rich deposit of fuel in the combustion chamber which will cause the motor to start much easier. The various control devices are shown in Fig. 57, while Fig. 58 shows the arrangement of the controls and instruments on the late models.

SPEED CONTROL

Clutch Pedal. There are three foot pedals that largely control the operation of the Ford Car. Of course, the throttle and the spark have a great deal to do with the speed of the car, but the foot pedals change the relation of the speed of the motor to that of the rear axle. The first pedal at the driver's left is for low and high speed, generally called the clutch pedal. When pressed forward, the clutch pedal engages the low-speed gears, causing the car to move very slowly but with great force, as the gear reduction is also great. This gear is also used when the car is traveling up a steep grade. When the clutch pedal is halfway forward, all gears are in neutral, being disconnected from the drive to the rear wheels, and when the hand lever is pulled halfway back, the clutch pedal will be held in the center, or neutral, position. When the clutch pedal is allowed to come all the way back —toward the driver—by pushing the hand lever forward, the clutch is thrown in, which causes the drive shaft to turn at the same speed as the motor. This is generally spoken of as direct drive, or high gear.

Reverse=Speed Pedal. The second, or center, pedal is used for reversing the motion of the car. When this pedal is depressed, the hand lever should be in the neutral position or, what amounts to the same thing, the clutch pedal should be held in the central position with the driver's left foot. The reverse pedal may then be depressed, which operation will cause the car to back up.

Brake Pedal. The right-hand pedal is used as a service brake, this brake being operated on the transmission drum; depressing the pedal causes the brakes to be applied.

Hand Lever. The purpose of the hand lever is to hold the clutch in the neutral position. If it were not for this lever, the driver would be compelled to stop the motor whenever he left the car. This lever also applies the emergency brakes at the brake drums on the rear wheels, thereby preventing the car from creeping forward when it is being cranked. The emergency brakes also hold the car when it is going up hill or standing at a curb or on an incline and are employed when it is desired to stop the car suddenly, etc. The brakes, however, are not operated until the lever is pulled back the entire distance. When the lever is in a halfway position, or almost vertical, the clutch is thrown out, and when it is placed all of the way forward the clutch is engaged, driving the drive shaft direct. When the car is to be reversed, this lever should be placed in a central position as this will prevent the clutch from dropping into high gear. See Fig. 107 for the position of these pedals.

Starting the Car. After the motor is started and it is intended to make the car move, the driver should gradually depress the clutch, or low-speed, pedal, thus bringing the low-speed gears into operation. It is best to throw the hand lever all the way forward, at the same time holding the low-speed pedal in the neutral position, before the low-speed pedal is depressed, as this operation will then eliminate any further movement of the hand lever until the car is stopped.

After the car has gained sufficient headway, say 20 or 30 feet, the throttle should be slightly closed and the foot removed from the clutch pedal, allowing it to come all the way back and engaging the clutch, thus causing the car to be operated in direct drive. The speed of the car is now controlled by opening or closing the throttle. The low-speed gear should never be used except when necessary, although it is not advisable to cause undue strain on the motor in order to prevent the low-speed gears from being used.

Stopping the Car. When the driver desires to stop the car, the high-speed clutch is released by pressing the clutch pedal forward to the neutral position. The foot brakes should then be slowly applied until the car comes to a dead stop. It is necessary to pull the hand lever in the neutral position before the driver removes his foot from the clutch pedal. If this is not done, the high-speed clutch will be engaged and the motor will stall.

Before stopping the motor, the throttle should be opened a little and then the ignition switch should be turned off. This allows the motor to stop with the cylinders full of fresh gas, thus enabling it to start very easily.

Spark lever. The spark lever is controlled by the left hand and should be placed in such a position that the engine will not knock; this position should be as far advanced as possible. If the spark is too far advanced, a dull knock will be heard in the motor. This knock is caused by the explosion occurring before the piston of the engine has completed its compression stroke. The very best results are obtained when the spark occurs at a position as far advanced as possible without knocking. The spark should be retarded only when the engine slows down on a heavy road or a steep grade. Care should be exercised not to retard the spark too far, since when the spark is late, a slow burning of the gas with excessive heat will result instead of getting the full power from the explosion.

The greatest economy in operation is obtained by driving with the spark advanced sufficiently to obtain the maximum speed with a given throttle opening. After a little experience, the driver will become accustomed to manipulating the spark automatically with excellent results.

Throttle. The throttle is controlled by the right hand and is used to increase the speed of the car to meet the various road and speed conditions. It is seldom necessary to use low gear except to give the car momentum in starting; therefore practically all the running speeds needed for ordinary travel may be obtained on high gear. The speed of the car may be temporarily slackened when driving through crowded streets and highways by slipping the clutch. This slipping is accomplished by partially depressing the high speed or the clutch pedal. This operation, however, should be used as little as possible as it causes excessive wear on the clutch plates.

LINCOLN SEDAN, SEVEN-PASSENGER

Started in Volume 1

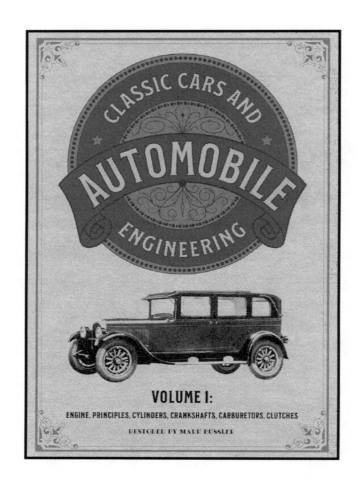

Continued in Volume 3

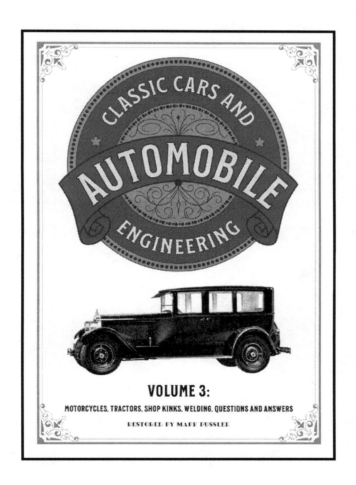

OTHER BOOKS FROM CGR PUBLISHING AT CGRPUBLISHING.COM

1939 New York World's Fair: The World of Tomorrow in Photographs

San Francisco 1915 World's Fair: The Panama-Pacific International Expo.

1904 St. Louis World's Fair: The Louisiana Purchase Exposition in Photographs

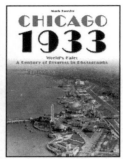

Chicago 1933 World's Fair: A Century of Progress in Photographs

19th Century New York: A Dramatic Collection of Images

The American Railway: The Trains, Railroads, and People Who Ran the Rails

The Aeroplane Speaks: Illustrated Historical Guide to Airplanes

The World's Fair of 1893 Ultra Massive Photographic Adventure Vol. 1

The World's Fair of 1893 Ultra Massive Photographic Adventure Vol. 2

The World's Fair of 1893 Ultra Massive Photographic Adventure Vol. 3

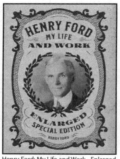

Henry Ford: My Life and Work - Enlarged Special Edition

Magnum Skywolf #1

Ethel the Cyborg Ninja Book 1

The Complete Ford Model T Guide: Enlarged Illustrated Special Edition

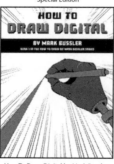

How To Draw Digital by Mark Bussler

Best of Gustave Doré Volume 1: Illustrations from History's Most Versatile...

OTHER BOOKS FROM CGR PUBLISHING AT CGRPUBLISHING.COM

Ultra Massive Video Game Console
Guide Volume 1

Ultra Massive Video Game Console
Guide Volume 2

Ultra Massive Video Game Console
Guide Volume 3

Ultra Massive Sega Genesis Guide

Antique Cars and Motor Vehicles:
Illustrated Guide to Operation...

Chicago's White City Cookbook

The Clock Book: A Detailed Illustrated
Collection of Classic Clocks

The Complete Book of Birds: Illustrated
Enlarged Special Edition

1901 Buffalo World's Fair: The Pan-
American Exposition in Photographs

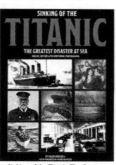

Sinking of the Titanic: The Greatest
Disaster at Sea

Gustave Doré's London: A Pilgrimage:
Retro Restored Special Edition

Milton's Paradise Lost: Gustave Doré
Retro Restored Edition

The Art of World War 1

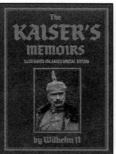

The Kaiser's Memoirs: Illustrated
Enlarged Special Edition

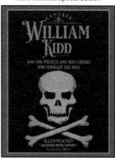

Captain William Kidd and the Pirates
and Buccaneers Who Ravaged the Seas

The Complete Butterfly Book: Enlarged
Illustrated Special Edition

Made in the USA
Monee, IL
11 September 2024

65388501R00234